Fulgence Marion, Maximilian Schele de Vere

The Wonders of Vegetation

Fulgence Marion, Maximilian Schele de Vere

The Wonders of Vegetation

ISBN/EAN: 9783337365592

Printed in Europe, USA, Canada, Australia, Japan

Cover: Foto ©berggeist007 / pixelio.de

More available books at **www.hansebooks.com**

ILLUSTRATED
LIBRARY
OF
WONDERS.

CHARLES SCRIBNER & CO: NEW-YORK.

VICTORIA REGINA.

THE
WONDERS OF VEGETATION.

FROM THE FRENCH OF
FULGENCE MARION.

EDITED, WITH NUMEROUS ADDITIONS,
By SCHELE DE VERE, D.D., LL.D.,
Of the University of **Virginia**.
AUTHOR OF "STUDIES IN ENGLISH," "AMERICANISMS," ETC.

WITH 61 ILLUSTRATIONS.

NEW YORK:
CHARLES SCRIBNER & CO.,
1872.

STEREOTYPED BY
WILLIAM McCREA & CO.,
NEWBURGH, N. Y.

RIVERSIDE, CAMBRIDGE:
PRINTED BY H. O. HOUGHTON AND COMPANY.

LIST OF ILLUSTRATIONS.

INTRODUCTORY NOTE.

THE zealous student who wishes to fathom all the mysteries of Botany must needs have scientific hand-books, drawn up with a careful regard to perfect accuracy, and containing in systematic order the outlines of his science. The general reader, and especially youths, on the other hand, require to be enticed by more attractive and popular works. The following pages are intended to be such an introduction to the Science of Botany. The author has selected some of the most wonderful plants of the vegetable kingdom, and the most remarkable phenomena connected with them, which cannot fail to interest. In issuing the work in English, care has been taken to secure scientific accuracy, and to bring the work up to a recent date.

THE WONDERS OF VEGETATION.

INTRODUCTION.

THE aim of this little work is to illustrate by characteristic and striking examples one of the aspects of the marvellous power of Nature. For Nature is neither as familiar nor as dear to us as she ought to be; and, as the tastes of society are daily becoming more artificial, we are likely to remove farther and farther from our great mother. It seems in fact as if that science which seeks to discover her secrets knows nowadays no higher aim than to apply these to the industries of man and perhaps to gratify curiosity. And yet it is only by intimate intercourse with nature that we can hope to extend our knowledge and to develope the affections of our heart. The more we aliena'e ourselves from her, the more we isolate ourselves and the lower we sink in intellectual greatness; while the closer we draw to her, the higher we rise in knowledge and in moral worth.

The magnificence and the glory of Nature may be studied in all her works and are manifested in even the smallest and apparently the most insignificant of all her productions. Without doubt the imposing

spectacle of the heavenly bodies moving in their or-
bits, and the wonderful forces brought into play to
control their motions, astonish us by their vastness
and their power; but the surprise awakened in us at
the view of the celestial wonders has its source chiefly in the comparative material greatness of the objects we contemplate. The Author of Nature mani-
fests His greatness as much in the germination of a
plant or in the generation of a living being as in the
guidance of a sun across the starry fields. It is His
Almighty Hand that studs the heavens with millions
of stars; but it is the same hand that daily scatters the
wind-blown seeds of earthly flowers upon the soft soil.
Both works reveal the action of an infinite intelligence. To rescue a world beaming with life from the
fiery fury of a comet, or to close a corolla at the approach of cold fogs or the touch of the north wind;
to spread out in space a milky-way rich in suns or to
adorn our garden trees with purple blossoms; to direct the gradual formation of the successive layers of
the earth's crust or to ripen the fruits that refresh us
in summer—these are equally the works of a Divine
Hand, a hand that recognizes no difference between
the small and the great.

To contemplate nature in flowers or in stars is
only to reach Truth in various ways: in both cases we
try to fathom the mysteries of the infinite in its different manifestations, to study the world under a
thousand aspects, to study Nature under two distinct
masters, but in the same school.

A full and complete description of the marvels of

vegetation would be a gigantic task; for as we have said, all the works of the Infinite are equally wonderful; all is marvellous in nature and the wonders of vegetation are limited only by the vegetable world itself. The most modest of plants in our fields, peeping modestly through the thick grass and those which reveal themselves under the microscope only, are quite as marvellous as the splendid orchid, the hoary cedar, the trembling sensitive plant and the poison tree. But in vegetation as in all things, the objects that appear to us really marvellous are those which awaken within us the most lively impressions. Owing to the natural inactivity of our mind custom has the effect of blunting our sensibilities, and of rendering our impressions less lively by familiarity; and thus the wonders which at first secure our keenest attention and awaken our most lively surprise, come in the course of time to be regarded with indifference. "Custom stales their infinite enchantment." What is unknown,—what is new —will always seem striking to us and secure our attention. In proportion as the objects become familiar they lose the power of exciting our wonder. Yet, strictly speaking, two objects of equal interest can never alter their relative position however accessible either may become on investigation.

Suppose one of us, living in a distant world, should reach the earth to-day for the first time—what would be his surprise at beholding around him all the manifold scenes, which, taken together, make up the great work of nature! The year is just beginning, and, as at the dawn of day, joyous Spring reawakens the

sleeping forces and once more decks with beauty the
world that has been stripped by the rough hand of
Winter. The heavens seem to have undergone a new
birth—their fresh blue kisses the distant horizon ; the
soft breeze plays with the swelling buds ; the sun pours
down from on high his vivifying rays, verdure springs
up anew, flowers and trees feel their life's blood once
more coursing through all their veins, and from the
highest mountains where vegetation still lingers, to
the green plains below, the whole earth seems to celebrate the return of Spring in joy and brightness.
What a marvellous transformation has been effected !

The trees of our orchards, the vast forests, which for
months have presented only bare trunks, and seemed
immovable, inert objects, which death had claimed
for its own, become green again, clothe themselves
once more with fresh leaves and spread their shade
over the peaceful retreats of the country. The habit
of seeing each year repeating the same marvel—the
same resurrection from death to life, keeps us from ap-
preciating it in all its grandeur, and recognizing in it
the prodigious forces at work. But if we think for an
instant of the aspect of winter and of that of the season
which succeeds it, we will be ourselves surprised at
the indifference with which we often regard these
changes, without even a passing glance or an earnest
thought.

How much more, if to the general contemplation of
the great transformations of spring and summer we
should add the special observations of each class of
plants. Then, if we attempted to follow the devel-

opment of each one of the various plants that grows on the face of the globe, the wonder would be proportionately greater. No two different species present the same phenomena, and from the time of the appearance of their first leaves, to the ripening of their fruits, they offer each a different spectacle.

Some plants guard their flowers from every glance, and seem averse even to showing their stems and leaves; others again, appear to have been created only for show, and display to the dazzled glance their wealth of sparkling and magnificent colors; still others seem to be of a more serious character, and, disdaining the frivolity of their gay companions, do not reveal their existence till the time when their ripened fruits attest their usefulness. Here the eye looks amazed at the undiminished vigor of an aged oak, which has seen in the time of our ancestors, processions of Druids pass through the gloomy forests, and which is so old that it forgets the number of its years, during which winds and tempests have in vain tried to uproot the colossal structure. There we behold a plant so fragile that it can hardly bear being touched, and resents the fluttering of a bright-colored bird by a painful shrinking of all its leaflets. But we have not yet spoken of the marvellous wealth of colors. What pencil can reproduce those various tints that adorn our beautiful flowers? In the meadows we trample under foot whole hosts of tiny blossoms which hide in the grass; purple petals overhang the banks of the stream whose murmur attracts us; at the roots of great protecting trees wild violets exhale their

sweet perfumes, but all the beauties of the world of
plants remain unappreciated by many.   They pass
by the dazzling white of the superb lily and bestow
not a glance; lovely rosebuds open to send forth
their splendor, unnoticed by the eye of man, and yet
can even the most perfect works of man's hand compare
in beauty with the most modest of the floral works
of nature ?

Nor are the splendid colors and the harmonious
tints of flowers their only charms ; even these beauties
are often surpassed by the rich perfumes of which
they preserve a rich treasury in their bosoms.   Does
it not seem then that flowers are the most opulent of
created things, that Nature has lavished upon them
her choicest gifts and that she loves them best of all
her children?   Well may we ask, as we inhale the
evening breeze, laden with rich perfumes, what strange
gifts they bring us from flower and forest and what
magic effect these grateful odors have on our mind,
on our soul?   They seem to be almost spiritual in
their ethereal lightness and to possess powers bestowed
upon them from on high.   They cannot be weighed
or measured, and we have as yet not succeeded in
fathoming their marvellous secret.

It is true then that all is marvellous in the vegetable world, and that in describing their marvels we
should be bound to describe every thing.   But since
it is equally true, as we have said above, that objects
which are continually before our eyes cease to interest
us, and since the new and the unknown alone appear
to us marvellous, we must needs seek among these

latter our illustrations of the wonders of vegetation. We shall, therefore, go beyond the narrow circle of our daily observation, and the facts which we are about to present will at least possess the charm of novelty as far as our daily thoughts are concerned; and if we can derive no interest from the things which surround us, we will go further afield. Travelling is a good master, let us follow him.

THE SACRED TREE OF THE ISLE DE FER.

The Pine of the Mountains.

PART FIRST.

CHAPTER I.

GENERAL VIEW OF THE GEOGRAPHICAL DISTRIBUTION OF
PLANTS ON THE SURFACE OF THE GLOBE.

THE carpet of plants which covers the earth does
not present throughout its whole extent a unity
of character independent of locality; on the contrary,
each climate has its own physiognomy, and certain
species seem to have a preference for certain countries.
Some delight in the burning soil of the tropics or develope their profuse wealth of fruit and flowers in the
warm and damp forests of the equator; others shun
the heat of the sun and prefer temperate zones or
even the lands of the North. It is this fact that

gives to each country its peculiar aspect. The animal world is too small in number, too restless and ever-changing to impress its mark on a country. Hence of the three kingdoms of nature the vegetable kingdom is that which has the greatest power in impressing on our mind an image of a land we have seen, and of the great natural divisions of the earth. For rocks and mountains also preserve the same forms everywhere from the equator to the poles, and their aspects could not give a special physiognomy to any country. But the trees and the flowers, the aspect of fields and meadows, of hills and plains, the forms and the colors of the leaves and the size of the plants—these give a character to the scene, on which we pass our lives and with which we feel bound up as if it were a part of our existence. And in long journeys amid the rich and abundant productions of the tropics, the traveller looks sadly and with regret for the trees of his native land; and his heart beats quicker as he sees at his feet a plant or a flower of his father-land recalling to him sweet memories of home.

The chief cause which rules in botanical geography and governs the distribution of plants throughout the countries of the globe, is temperature. Thus here also, as in the whole harmonious life on earth, the sun reigns as a sovereign—it is he who directs the orchestra, calling forth now soft and solemn cadences, now light and brilliant melodies. Two hundred thousand varieties of plants divide the surface of the earth among themselves. One great law directs the division, the law of temperature. No other force exercises any thing like the degree of influence which this agency brings to bear upon the distribution of plants.

Let us for an instant consider the world as a sphere turning upon its axis—an ideal line passing through its centre—and call the two points where this line would reach the surface as respectively the North and South poles. The motion at the poles will of course be inappreciable. We will give the name of the equator to the great ideal circle which passes round the middle of the earth, dividing it into two equal halves —the North and South hemispheres. Now, as the rays of the sun are so much the more oblique the more they diverge from the equator, it follows that the heat at the equator is at the maximum while at the poles it is at the minimum. With the decrease of heat from the equator toward the poles corresponds the geographical distribution of plants. At the equator and in the neighboring tropical regions we meet with the vast proportions of the largest plants—as the boababs, the palms, the elegant tree-ferns, the aloes, the heaths— magnificent plants which love and seek heat. In leaving these heated countries we encounter olives, laurels, mimosas and bamboos. Continuing our route towards the poles we see magnolias, chestnut-trees, cotton plants and witchelms. Proceeding still further from the tropics till we reach the latitude of France and Middle Europe, or our Middle and Eastern States, we meet with oaks, beeches, willows and elms, with our common fruit-trees and our cereals. In the Northern countries, near the limits of vegetation, we

2

still find the mountain ash, the pine-tree and all the conifers.

At last even these plants cease to grow—the oak, the hazel and the poplar at 60° North latitude; the birch and lime at 63°; and the conifers are not found higher than 67°. Beyond 70° only a few stunted willows grow here and there close to the ground. Further north, beyond 75°, not a tree is to be found, shrubs and plants even have disappeared and cereals can no longer exist, for even barley and oats are not found beyond 70°.

The local physiognomy of plants depends thus upon the normal temperature of each climate. The same principle is applied to the elevation at which plants grow, and by combining the two, we are enabled to understand in its entirety the distribution of plants over our globe.

Instead of travelling from the equator towards the poles, let us simply ascend a high mountain and we will find that the classes of plants appear in the same order, following the thermometric ladder of altitudes. We know that the higher we rise in the atmosphere, the lower we find the temperature; and this fall is so rapid that the ascent of a few minutes in a balloon, or of a few hours upon a mountain side, takes us through all the different degrees of temperature from 70° or 80° of heat on the plains to the freezing-point, and below in the higher regions of the atmosphere. In consequence of this law all the mountains of the globe have a lower temperature at their summit than at their base; and we find among their vegetable productions the same characteristic zones of vegetation which we meet with in going from the equator toward the North or South pole. We might therefore compare the two terrestrial hemispheres to two *mountains, supported the one against the other by their bases at the circle of the equator,* their summits covered with eternal snows, and certain classes of plants dwelling on their sides in regular succession as we move from their tropical base to their polar summit.

We can obtain an idea of the succession of plants by following an ascent made by Mr. Ch. Martins, of Montpellier, who divides with Humboldt, Hooker and a few renowned botanists the glory of advancing the geography of plants—a science which had its birth at the commencement of the present century. The following are the observations made by him during the ascent of Mount Ventoux in Provence:

"All the trees belonging to the lowest plains," he says, "were found at the base of the mountain, the characteristic trees being the Aleppo pine and the olive-tree. The first does not appear at a greater elevation than 1,300 feet above the level of the sea, the second ascends farther, to about 1,500 feet. Besides these trees we saw all the Southern species which characterize the vegetation of Provence, the kermes oak, the rosemary, and the Spanish broom. A narrow zone succeeded next, the chief feature of which was the evergreen oak, which is not found beyond 1700 feet.

" A treeless region comes immediately after the
first two. The soil is bare, stony, and generally un-
cultivated. Yet here and there we saw some fields
of chick peas, of oats or of rye—the last being found
at 3,100 feet above the level of the Mediterranean;
only one shrub, the box, thyme and lavender, and a
few other plants, shared with each other these desert-
ed regions. Beeches were seen at 4,000 feet. At
this height there is very little shelter to be found, and
the trees, exposed to the strong action of the wind
which bends them down to the ground, are really no
larger than bushes.

" At the height of 5,100 feet the cold is too keen,
the summer too short, and the wind too violent to allow even the beech any longer to thrive. On the
Ventoux, as well as on the Alps and the Pyrenees, a
coniferous tree is the last representative of arborescent plants. It is a very humble kind of pine, called
the 'mountain pine.' These pines rise to the height
of some 10 or 15 feet in a few sheltered places, but become thick shrubs where they are exposed to the
wind. They are found at an elevation of 5,430
feet, and mark the extreme limit of arborescent vegetation.

" Thus the plants teach us, as well as the best ba-
rometer, that we have reached the region where trees
disappear, but where the botanist will find, with delight, the plants of Lapland, of Iceland and of Spitzbergen. In the Alps, this region extends to the limits of perpetual snow, the home of eternal winter;
but as the Ventoux rises only to the height of 5,800

feet, its summit corresponds only to the lower part
of the Alpine region, in the Alps and Pyrenees.
At this height trees have entirely disappeared; but
a number of small plants unfold their petals on the
surface of the stones or rocks. These are the or-
ange-flowered poppy, the violet of Mount Cenis, the
blue-flowering astragal, and at the very summit the
meadow-grass of the Alps, Gérard Euphorbia, and the
common nettle, which appears wherever man builds
a house.

"On the northern slope we find the saxifrage,
which inhabits the Alpine summits amid perpetual
snows, and covers the icy shores of Spitzbergen."

Thus, whether we travel from the hot countries of
the equator to the icy regions of the pole, or rise
from temperate plains to snowy mountain summits,
we find that the heat-giving power of the sun alone
governs the distribution of plants. Each plant seeks
its own degree of heat. The dwarf birch resists a cold
of 40° below zero, the orchids are frozen at 50°
above it. On the other hand, each species requires a
certain amount of heat for germination; and after
that, additional heat is necessary to enable it to flower and ripen its seed. Our precious cereal, wheat,
refuses to ripen its golden ears, rich with bread that
makes the strength of man, unless it has 3600° of heat
in all, for the time from its first springing up till harvest; and the darkling cluster of grapes cannot glad-
den the heart of man with its red wine, unless it have
5400°. This is why each plant shows a decided preference for a particular locality and a particular temper-

ature, why crops vary in different years according to
the amount of heat, and why each region of the globe
presents a peculiar physiognomy in its plants which
corresponds with the average degree of heat that there
prevails.

FOREST OF MANGROVES.

Vegetation in the Tropics.

CHAPTER II.

VEGETATION IN THE TROPICS.

IN order to obtain some approximate idea of the grandeur and the magnificence of the vegetable world, we must leave the temperate climates to which we may have been accustomed, and pass from under the cold northern sky to the countries that are loved by the sun, where nature still lives in all its youthful vigor, and glows in full luxuriance; where the earth preserves, as it were, a living museum of all the riches which have elsewhere disappeared in the immense succession of primitive ages. For this purpose we will follow certain travellers upon whose reflections— on the power of nature as manifested in the plants of the tropics—both science and poetry have had their influence.

"Vegetation displays the most majestic forms under the fiery rays which flood down from the tropical heavens," says Humboldt in his "Pictures of Nature." In the land of palms, in place of the meagre lichens and mosses of the North, we have the cymbidium and the fragrant vanilla hanging from the trunks of the cashew nut and gigantic fig-trees. The fresh verdure of the dracontium and the deeply indented leaves of the pothos contrast with the brilliant colors of the orchids. The creeping bauhinia, the passion flower, the yellow banisterias, interlace the trees of the forest and throw their trailers far into the air. Delicate flowers spring from the roots of the theobroma and from the rough bark of the crescentia and the gustavia. In the midst of the luxuriant vegetation, among the confusion of creeping plants, the Naturalist has often difficulty in determining to what stem the flowers and leaves before him belong. A single tree interlaced by the paullinia, the bignonia and the dendrobium, forms a group of plants which, if separated, would suffice to cover a considerable space of ground.

" Tropical plants contain a great deal more sap, and their leaves are much larger and more brilliant than those of the North. The plants which minister to the household wants of man, and render our vegetation so uniform, do not form a feature in tropical vegetation, which is consequently much more varied than ours. Trees nearly twice the height of our oaks bear flowers which equal our lilies in size and in brilliancy. Upon the umbrageous banks of the Rio Magdalena in South America, grows a creeping aristolochia, the flowers of which are four feet in circumference, so that the children amuse themselves by making head-dresses of the blossoms. The flower of the rafflesia is nearly three feet in diameter and weighs nearly fifteen pounds.

"The extraordinary height to which not only the mountains but whole countries rise at the equator, and the depression of the temperature, which is the result of that elevation, enables the inhabitants of the Torrid Zone to behold an extraordinary spectacle. At the same moment that they see around them the palms and bananas of the South, they are called upon to notice a number of vegetable forms which ordinarily belong only to northern lands. The cypress, the fir and the oak, the thorn and alders, very much like our own, cover the plateaux of Southern Mexico and the part of the Andes which crosses the equator. Thus nature permits the inhabitant of the Torrid Zone to see growing near each other all the vegetable forms of the earth without leaving the place where he was born, just as the vault of heaven displays to his view all the world of life from the one pole to the other. These enjoyments and many others are denied to the son of the North. He never sees a large number of the stars, nor does he ever behold many of the most beautiful vegetable forms, such as the palms, the tree ferns, the bananas, and the mimosas with their delicate feathery leaves.

"The few sickly exotics which we raise in our green-houses represent the majesty of tropical vegetation only very imperfectly; but we find abundant compensation in the beautiful language and the brilliant imagination of the poet, and in the imitative art of the painter which enables us to create a tropical world of our own and pass in review before our mind the living forms of exotic nature. In the cold climates of the North, in the midst of sterile plains, man can appropriate to himself the labors of others and enjoy at home what the traveller has gone far to seek."

To this sketch, taken from one of the grand founders of the science of the geography of plants, we will add a few sentences from the gifted and painstaking author of "Scenes of Nature under the Tropics," which are worthy to be placed by the side of the words of the great master.

"Upon the banks of the lakes and the rivers," says Denis, "the heat of the sun, calling into activity the beneficent moisture of these vast reservoirs, produces gigantic forms of vegetation. Trees which elsewhere grow with difficulty, rise here majestically and embellish the banks at the same time that they attest their fertility. The Amazon, the Ganges, the Niger roll their waters through vast forests which, being replaced from age to age by new growth, have always resisted the efforts of man. It seems indeed that Nature chooses the banks of these immense rivers to display here a magnificence unknown in other places. I have noticed in South America, that the trees, rising to an immense height near the rivers, give a peculiar aspect to the forests. Not that in such places Nature presents an appearance of absolute disorder; on the contrary,

SOUTH AMERICAN SCENE.

it seems as if its strength and its grandeur have specially enabled it here to display a certain majestic regularity in vegetation. The trees, towering up to a
height that wearies the eyes, do not permit feeble
shrubs to grow underneath. But the vault of the forests is raised higher; the enormous trunks of the trees
which support it form immense porticoes and spread
out their branches with majesty. They are covered
at the top with a multitude of parasitical plants, which
seem to claim the air as their domain, and which
proudly mingle their flowers with the very top branches. Here often upon the immense fig-tree, which is
itself unpretending in appearance, a flexible liane will
twist spirally around it, covering it with garlands,
and uniting it to all the great plants that grow around,
till at the utmost top it seems to defy the dazzling
splendor of the noon-tide before it once more descends
to embellish the mysterious recesses from which it
first sprang."

In the forests, which are less majestic and more
easily penetrated by the rays of the sun, vegetation
presents an astonishing variety and within much easi-
er reach. Among the travellers that have described
these forests in detail, perhaps no one is more exact
than the prince of Neu Wied.

"Everywhere life and vegetation abound without
limits," he says; "and not the smallest space can be
found where there are no plants. On all the trunks
of the trees we see grenadillas, caladineas, pepper
and vanillas, etc., flourishing, climbing and twisting.
Some of the gigantic stems covered with flowers look

at a distance white, deep yellow, bright scarlet, rose colored, violet or purple, even sky blue. In the marshy places rise in compact groups the large and beautiful elliptical leaves of the heliconia, which are sometimes eight or ten feet in height, and bear for an ornament extraordinary flowers of deep red or flame color. Enormous bromelias, with countless flowers, hold huge trees in deadly embrace, till they die after a long struggle and suddenly fall thundering to the ground. Thousands of creeping plants, from the smallest size to the thickness of a man's thigh, with a hard and compact wood, twist around trees, rise to their summit and there flower and bear their fruit without ever being seen by human eye. Some of these plants, like certain banisteria, have forms so singular that we cannot behold them without astonishment. Sometimes the trunk around which these plants have twined themselves dies and falls away in dust. Then the huge stems of the parasites, strongly interlaced, are seen supporting each other, clearly showing their frail mutual support. It would be very difficult to present a faithful picture of these forests; for it is not within the resources of art to represent them as they are."

There is in the forests of the New World a harmony perfectly in accord with the phenomena presented to the view—as all is grand, imposing and majestic; the songs of the birds and the cries of the different animals also have something savage and melancholy in their utterance. Brilliant and sustained cadences, cheerful chirpings, lively and gay modulations, such as we hear in temperate zones, are here less frequent—they are replaced by songs more grave and measured. Now a voice is heard which seems to imitate the far-sounding blow of the hammer upon the anvil, and now a sound falls upon the ear which resembles the sudden breaking of the strings of a violin. All over the forest you hear strange sounds which cause profound astonishment; but often at sunset, when the birds have ceased their songs, there is heard from the highest tree-tops a voice which would fill the traveller with fear if he were ignorant of the cause. Murmurs like those of the human voice announce that the guaritas (*Simia Beelzebub*) are beginning one of their assemblies, which are said to be held in honor of the setting sun. Their howls, prolonged in the most lugubrious manner, have caused credulous men to believe that these animals are rendering homage to Satan and paying him a tribute which he exacts. These sounds, heard at the hour when the day dies, are solemn and imposing—they give a character of sadness to the scene. If the jaguar and the black tiger roar, they fill the forest with a sound which is majestic, but productive also of uneasiness. Harmless animals hearing this dreaded voice suddenly become silent, as if they feared to mingle their utterances with those of the awful master. If in addition to these sounds the wind begins to blow violently, bending the lofty summits of the trees, making the palms sigh as they bend low and mingle their moans with the rustling of the lianas, and losing itself finally in the sombre depths of the primitive forests, then the sounds become so sad and mournful that admiration gives way to terror.

Of the great plants which attract the attention of the traveller and impart to the vegetation of the tropics an altogether foreign aspect in the eyes of the foreigner, we will select the most remarkable, whether from their beauty and size or from the service which they are made to render to the natives. The latter aspect will be of special value to us, for it will give us an idea of the power and spontaneousness with which nature here proceeds in her work—supplies all that is needed, and incessantly gives new beauty to life. To mention only one example in direct connection with the descriptions which follow, we will remind the reader that if plants and animals are the natural food of man, that food must needs vary according to the countries which he inhabits. Where a certain mode of life is no longer possible because of climate and soil, that mode of life is changed; but life itself is not suspended on that account. To maintain life is the supreme aim of all the forces of Nature, and her law is to manifest herself under all possible forms. In northern countries the cereals, and wheat and corn in particular, supply our daily bread, while wine, beer, cider are the drinks according to the various countries. But in order that the wheat may ripen there must be frost during the winter; otherwise it grows rank and bears no seed. Now in warm countries there is no winter; the seasons, distinctly marked in northern latitudes, become effaced in proportion as we approach the equator, and

FOREST IN BRAZIL.

in the tropics neither wheat nor any other cereal will
thrive; but are these countries on that account uninhabitable? By no means. When wheat no longer
ripens, other vegetables take its place; trees furnish
our daily bread and wine in their fruit; milk flows
in the shape of creamy, sap and the fruits of northern
countries are replaced by those of another climate.
Let us select the characteristic types of these valuable
plants, and if we cannot see them in their own country let us at least call them up before our mind's eye
and make them tell us their own history.

Bread-Tree of Tahiti.

CHAPTER III.

THE BREAD-FRUIT-TREE.

WE will commence this portion of our task by mentioning certain extraordinary plants, which in countries essentially different from our own in soil and climate, are made use of by the natives to supply those wants which are supplied among us by means of certain domestic animals, or by mechanical contrivances, put into daily operation. There is one remarkable family in the kingdom of Flora, of which some members furnish leavened bread, others a supply of milk, equal to the best cow-milk, and still others the most fearful poison as yet known to man. The useful members are the bread-fruit-tree, the milk-tree,

and those which supply limpid water or some strengthening beverage to the traveller.

As bread is the staff of life, we will give the place
of honor to a fig-tree, which actually grows bread for
our antipodes in Oceanica, and thus renders unneces-
sary the toils of the sower, the reaper, the miller, and
the baker.

The ancients loved to consider Nature as an individual being, apart from the world, endowed with
reason and will, and constantly spoke of her in prose
and poetry as the " Universal Mother," and she well
deserves this beautiful name, by her conduct toward all
living things, and especially by the motherly affection
with which she provides for the numberless children
to whom she is incessantly opening the gates of existence. For what else are the rays of the sun calling
forth life upon the hill-slopes; the rain falling softly
on meadow and prairie, and even the warm carpet of
snow which winter spreads over the frozen earth;
the dew of morning and the vapory mists of evening
—what are they but so many evidences of the tender-
ness of our mother Nature—or rather the watchful-
ness of Divine Providence. But apart from these
cares bestowed impartially and without distinction
upon all existing things, the philosophical traveller
discovers, every now and then, special instances which
reveal to us more pointedly this marvellous goodness
of Providence, than the general working of the ab-
stract laws of nature.

Among the examples which in a special degree
attest the watchful care of Providence, we have to

mention that of the bread-tree, discovered in the isle of Oceanica. This invaluable tree belongs to the genus *Artocarpus*, of the fig family. The leaves in this family are simple, plain or serrated, and the flowers very small and imperfect, some having no corolla, and others no calix, but all appearing alike upon the same tree at the extremities of the branches.

The true bread-tree has indented or serrated leaves. We say the true bread-tree, for this genius embraces many other species, which, in spite of a very remarkable organization, do not possess the properties of the one we have mentioned. Thus there is an *Artocarpus incisa*, with small leaves and flowers, but bearing fruits which are, perhaps, the largest borne by any tree on earth. These round fruits are sometimes so large that a man cannot lift them! The kernels are eaten, roasted like chestnuts, but they are not easily digestible. Then there is the Jack (*Artocarpus integrifolia*), of the Indian Archipelago, with a huge trunk, and dense foliage on the broad-branching summit, while the fruit measures 18 inches by 15. Travellers are not agreed as to the merits of the latter. Rheede says they have an agreeable taste and odor, but Commerson could not summon courage even to put a morsel of it in his mouth. "Tastes differ," but it seems difficult to explain such contradictory opinions, unless it should be that these travellers speak of such trees as certain critics are said to judge of works which they have never seen. A third species is the *Artocarpus hirsuta*, the tallest of the genus. Its wood is used in carpentery, and in boat-building.

BREAD-FRUIT.

The Indians hollow out the trunk to make their pi-
raguas, some of which measure 80 feet in length by
nine in width, and thus enable them to make long
ocean voyages.

We return to the true bread-fruit-tree. The discoveries in Oceanica have rendered it celebrated, and
special expeditions have been undertaken for the purpose of obtaining roots for transplantation in different
parts of the Old and New World. We shall presently
notice the most remarkable of these expeditions. The
following are the distinctive characteristics of this tree :

The trunk is straight, as thick as a man's body,
and rises in a gentle spiral to the height of about 40 feet.
Its large round top covers with its shadow a space 30
feet in diameter. The wood is yellowish, soft and
light. The leaves, 1½ feet long and one foot wide, large
and permeated with seven or eight lobes, a form which
characterizes this species. The same branch bears
male and female flowers. The bread obtained from
the tree is its globular fruit, larger than a child's
head, weighing three to four pounds, and rough on the
outside, covered with hair. The thick green rind en-
closes a pulp, which, during the month that precedes
maturity, is white, farinaceous, and slighly fibrous ;
but when ripe, changes in color and consistency, and
becomes yellow and succulent or gelatinous. The
island of Otaheiti abounds in the best kind of these
trees, which bear fruit without seed ; the other islands
of Oceanica produce varieties of less valuable bread-
fruit, containing angular seeds almost as large as
chestnuts.

The fruit of this tree ripens during eight consecutive months in the year. The islanders live upon it as we do upon our manufactured bread—it is their main food, and nature as we see furnishes it to them without their being put to the trouble of cultivating the ground, of sowing, reaping, threshing, grinding or baking. To have their " fresh bread " they choose the time when the pulp is farinaceous, which they can tell by the green color of the rind. The necessary preparation " for the table " is accomplished by cutting them in thick slices and cooking them upon a charcoal fire ; when ready, each " loaf " weighs about a pound. They are sometimes also placed upon a heated oven, as we do with pastry, and left there until the rind begins to blacken. Then the burnt part is scraped clean, as your toast, and the interior is white, ready to be eaten, tender as the crumbs of French rolls, but little differing in taste from wheaten bread, except only a slight flavor suggestive of the inside of an artichoke. As the natives want bread throughout the whole year, they take advantage of the time when the fruits are abundant and prepare from the pulp of the surplus fruit a paste which, after being fermented, can be kept a long time without turning sour. During the four months when the trees do not yield, the natives live upon this preparation.

The expedition to which we referred was that made by Captain Bligh, sent in search of the bread-tree of Otaheiti for the purpose of introducing it into the tropical colonies of Great Britain to furnish food for the slaves.

The narratives of Cook and other explorers had encouraged the highest expectations of the benefits which would result from the culture of the bread-fruit-tree. The English colonists entreated their government to obtain for them this wonderful tree; a vessel specially fitted for the purpose was got ready and placed under the command of Bligh, then only a lieutenant, but afterwards an admiral. The selection of the commander was judicious; for Bligh had accompanied Cook in his voyages and given on many occasions proofs of his talents and his gallantry. Leaving England in 1787, the expedition arrived in six months at Otaheiti. The islanders received them hospitably; more than a thousand plants were put in pots and boxes and taken on board, with a sufficient quantity of fresh water to keep them alive, and five months afterwards the precious cargo was floating towards its destination. But in spite of all the happy auspices under which the return voyage was begun, it had an unfortunate ending. It furnished one of those examples, happily rare, of the revolt of a crew and the desperate position of a captain left to the mercy of the mutineers in the midst of the silent ocean. Twenty-two days after they had left Otaheiti the greater part of the crew having joined in a most cowardly plot, seized Bligh during the night and placed him with the eighteen that remained faithful to him in a long boat with some provisions and instruments, and leaving them alone in the middle of the

ocean, sailed off and were soon out of sight. Bligh
and his companions bore up with superhuman courage in the midst of their fatigue and sufferings;
only one succumbed. They arrived at the island of
Timor after having sailed the distance of 3,600 nau-
tical miles in the long boat. The Dutch governor
received them kindly, and soon twelve of them were
able to take passage to Ireland. Bligh obtained justice
in England; he was immediately promoted to the
rank of captain and placed in charge of a new and
larger expedition. This time he succeeded completely,
and two years after the two vessels of the expedition
landed in the British West Indies, having on board
1,200 plants of the bread-fruit-tree, and without having lost a single man of either of the crews.

The slaves of the West Indies did not show as
much alacrity in making use of the fruit as had been
expected, preferring their familiar food, the banana;
on the other hand, the Europeans accepted it with great
pleasure. It ought to be stated, however, that the
slaves eat the fruit without having previously prepared it, while the Europeans cooked it according to
the best receipts of English writers.

The old people of Otaheiti attribute the origin of
the bread-fruit-tree to an incident which is embodied
in a touching legend.

At a time of great scarcity, a father assembled his
numerous children upon the mountains and said to
them: "You will inter me in this place; but you will
find me again on the morrow."

The children obeyed, and coming on the following

day as they had been commanded, they were much
surprised to see that the body of their father had been
transformed into a great tree. His toes had stretched
out to form the roots; his powerful and robust body
had furnished the trunk, his outstretched arms were
changed into branches, and his hands into leaves. His
bald head finally had disappeared, and a delicious fruit
was found in its place.

This legend recalls the seventh circle of the Infer-
no of Dante, where the souls who had béen violent
upon earth are seen changed into living trees, while
their limbs writhe and twist like the branches of dead
trees. But we prefer the simple legend of the primitive isles to the gloomy imagination of the great Italian; the poet speaks of the dead, the islanders appeal
to the living.

THE MILK TREE.

Ever since the discovery of the New World by
Columbus, explorers have been hard at work to become
familiar with the new countries which were opened
up before them, and to publish descriptions of the new
forms of life both in the animal and the vegetable
kingdom. If we were to believe all the marvellous
narratives of the early times, from Marco Polo to Ma-
gellan, we might easily place in our Book of Wonders
men with dog-heads and trees gifted with the powers
of speech. But we do not mean here to repeat those
fables; we are interested only in natural and actual
wonders. As early as the year 1505 many remarkable
plants and animals had already been described truth-

fully, but some rare species were for a long time over-
looked, though they belong to the countries first discovered and though they ought to have attracted attention by the special features by which they are distinguished. To this number belongs the cow-tree,
the *arbre à lait* (milk-tree) of the French. This tree
is one of the most remarkable of Central America, and
yet it was not known to Europe as late as the beginning of the present century. It was on the 1st of
March, 1800, that Humboldt and Bonpland observed it
on the Barbula Farm during their expedition to the
valleys of Aragua.

Cow-Tree.

An ancient writer, Laet, had mentioned it briefly
in his *Novus orbis:* "In the province of Cumana,"
says he, "there are trees which, when their bark is
pierced, pour out an aromatic resin; while others yield
a juice, which resembles curdled milk, fit for food."
This solitary observation is obviously very incom-

plete, and it was left for Humboldt to describe the tree.
We quote the passage :—

" While these trees present the remarkable feature
of furnishing man with his bread all ready made by
the benevolent hand of Nature, he seeks his daily
bread in other less favored regions in every plant
that grows around him. From the lofty bread-tree
through the whole scale of trees and shrubs and lowly
plants, there are but few which he has not learned to
convert into nourishing food; and when at last the
earth seems to refuse him further aid, he digs beneath
the surface or dives into the water to bring back with
him the desired supply. One of the most remarkable
plants of the latter class is the Edible Arum (*Colo-
casia Esculenta*), the roots of which, if properly prepared, furnish a palatable dish. But the plant has
another and even more striking peculiarity. It literally distils water and launches tiny drops in the form
of a jet from the pores at the end of its magnificent,
heart-shaped leaves. A careful and ingenious observer
ascertained that from 10 to 100 drops of water were
thrown every minute to a distance of an inch and
more."

" In returning from Porto Cabello we rested at
the Barbula plantation. We had heard for some
weeks about a tree the juice of which was milk fit
for food. The tree was called Palo de Vaca, and we
were assured that the negroes of the plantation largely used this vegetable milk, regarding it as healthy
and nutritious. As all milky juices of plants are
acrid, bitter, and more or less poisonous, this assertion

seemed to us very extraordinary. Experience has however proved that the virtues of the cow-tree had not been overrated. When incisions are made in the trunk of the tree, it gives forth a glutinous milk, rather thick, free from all acidity, and exhaling a very agreeable odor. We were offered some of it in cala-bashes, and drank considerable draughts of it both that night before retiring to rest, and early in the morning, without experiencing any unpleasant effects. The viscous quality of the milk was the only thing un-pleasant about it. The negroes and free men who work on the plantations drink it, soaking in it corn cakes and cassava. The manager of the farm assured us that the slaves became sensibly fatter during the season when the cow-tree furnishes them the largest supply of milk."

"Among the numerous interesting phenomena presented to me during my expedition," continues Humboldt, "few things made a more vivid impres-sion upon my imagination than the appearance of the cow-tree. Every thing that relates to milk or concerns cereals awakens within us an interest which is not merely of a scientific character, but which connects itself with another order of ideas and sentiments. It is hardly possible to conceive of the human race exist-ing without farinaceous substances, or without that nutritious liquid which springs from the mother's breasts, and which is so admirably suited to the infant in the weakness of its early youth. Farinaceous mat-ter is found not only in grain but also in many roots, and even in the trunks of certain trees, as in the sago

MILK TREE OF GUIANA.

palm. But as to milk, we are accustomed to look upon it as solely an animal product. Such are the impressions which we have received from our infancy, and this was the source of the astonishment which seized us at the first sight of the cow-tree.

In Caracas, South America, there grows upon the dry face of a rock a tree the leaves of which are dry and barklike; its great roots penetrate with difficulty into the earth. During many months of the year not a shower moistens its foliage—the branches appear dead and withered, but when the trunk is pierced a sweet and nourishing milk flows forth. The supply of the liquid is most abundant at sunrise. At that hour the blacks and the natives, furnished with large pitchers to receive the milk, which is yellow and gradually thickens on the surface, arrive at the cow-trees from all quarters. Some drink their supply on the spot, others carry it away to their children. We might fancy that we beheld the family of a patriarch who is distributing the milk of his herd.

The milk-yielding plants belong, mainly, to three families of Euphorbiaceæ, Urticeæ and Apocyneæ, but in the juice of almost all of these acrid and deleterious elements are to be found, from which the milk of the cow-tree is free. Still there are some species of Euphorbia and Asclepias which also yield milk that is sweet and harmless. Thus in the Canaries we find the *Tabaïba* (*Euphorbia balsamifera*) mentioned by Pliny as *Ferula*, and as giving out, when pressed, a liquor agreeable to the taste; at Ceylon is found a lactiferous Asclepias, the milk of which is used when cow's milk cannot be had, while the leaves are used in cooking such food as is elsewhere prepared with cow's milk.

This natural vegetable milk offers besides other points of affinity and resemblance to natural milk. Thus, placed in the open air, in a short time a yellowish, thick membrane appears on the surface, not unlike the little skin that forms on milk, and this continues to thicken and is taken off, to be kept under the name of cheese often for a week. Nature, however, also takes to churning herself occasionally. On the banks of the Niger, the natives gather their butter directly from a tree (*Pentadesma butyracea*) and sell it in their markets. It is said that the kings of Dahomey, fearing its value as an article of export, and thus as a means of bringing the land into relations with more civilized countries, have ordered it to be exterminated! It is annually burnt by royal decree—it annually springs up again defying the decrees of the cruel sovereigns.

Although many kinds of lactiferous plants furnish caoutchouc, not a trace of it is found in the product of the cow-tree; and the cheese of which we have spoken are not very different from our own. In chemical analysis the tree milk bears a close affinity to animal milk; the butter is represented in the vegetable milk by a beautiful and abundant wax, caseine, by a substance not unlike the fibrine of blood, and the serum by a watery liquid containing a little sugar and a small percentage of the salt of magnesia. Placed over the fire vegetable milk undergoes the same modification as animal milk. A cream forms on the surface which cannot easily be taken off; the milk boils up and shows a tendency to run over from the vessel which contains it. If the cream is removed as it forms, and a steady heat is kept up, the milk gradually assumes the consistency of paste; then appear upon the surface oily rings, like those which come to the surface of cream that has been upon the fire for some time. Finally, this fat part envelopes the whole of the posset, which then diffuses an odor exactly similar to that of roast beef.

The tree is found chiefly in the valleys of Caucagua, in the Cordilleras, near the sea-coast, and in the vicinity of the lake of Valencia. At Caucagua the natives name it the "Arbo de leche" (milk-tree), and assert that they can tell by the color and the thickness of the leaves which trees contain the juice in the greatest abundance, precisely as the farmer tells by certain marks the good qualities of a milch cow.

In 1829, Smith, the traveller, while passing through the woods of Guiana, made special search for the tree which had been described by Humboldt, and inquired of all the guides if they could tell him any thing of such a tree. He had already met milk-giving plants, but the bitter taste of their sap was utterly unlike milk. One day finding himself in a little Indian village near the first rapids of the River Demerary, he heard reports of a tree called the hya-hya, whose milk, it was said, was nourishing and agreeable to the taste. Determined to ascertain the fact, the traveller sent an Indian in search of one of these trees. The Indian not only found the tree, but cut it down, and the creek, across which it had fallen, was whitened by its milk. A knife stuck into the bark immediately brought forth a stream of sap, which the Indian drank eagerly. Smith drank after the Indian, and found the milk excellent. "It was," he says, "thicker and richer than cow-milk, entirely free from bitterness, and the only slightly unpleasant feature was, that after drinking my lips felt slightly viscous. As I passed the night in the village," the traveller continues, "I had in the morning a glass of this milk for my coffee, and it proved such a good substitute for cow-milk that no one could have told the difference ; for the slight viscousness which I had noticed in tasting it before, disappeared when it was mixed with the coffee."

The milk flows more freely if the opening made is transverse or oblique, than if it is longitudinal. The bark of the hya-hya is gray, rather rough, and has to be cut completely through in order to make the milk flow. This tree is very different from the other cow-tree. Its leaves are elliptical, and grow in couples. The chemical composition of the milk also differs—it is not equally nourishing.

Besides these remarkable species of milk-producing trees belonging to America, a milk-tree not less remarkable, called *Masaranduba*, by the Indians, is found in the port of Para (Brazil). It is one of the largest trees in the Brazilian forests, and furnishes a wood highly prized for ship-building. The tree blooms in February and yields a delicious fruit, the taste of which recalls that of strawberries eaten with fresh cream. From an incision made in the trunk a white milk pours forth, perfectly liquid, agreeable to the taste and without any odor. The natives habitually live upon it. The officers of the Chanticleer, whose surgeon, Mr. Webster, was the first to discover this tree, constantly used it during their stay in the ports, like ordinary milk, both in tea and coffee.

This tree is very tall; its bark is dark brown and the leaves are large and oval.

The crew of the Chanticleer having kept some of this milk in bottles, found that at the end of two months it had separated into two parts, the one liquid, light yellow, and with a slightly sour odor, the other solid, white and insipid, insoluble in water or in alcohol. This substance would burn easily, giving forth a brilliant green flame; it appeared to consist for the most part of wax, and contained none of the animal matter which abounds to such an extent in the coagulated parts of the milk of the *palo de vaca* or *cow-tree.*

This milk-bearing tree, at first and still quite commonly called *Galactodendron dulce,* is in reality the same tree as the *palo de vaca,* now known by its botanical name of Brosimum Galactodendron, and belongs to the family of fig-trees. There are, however, on the coast several other trees which give forth milky juices and are often confounded with the latter. For instance, in the neighborhood of Maracaibo, the *Clusia Galactodendron* pours forth a milky stream very agreeable to the taste; but the milk is not of equally good quality, cannot be as easily purified, and leaves an unpleasant resinous matter behind instead of wax; still others, like the *Hura Crepitans,* are actually poisonous, and the sap is employed to poison the waters of rivers for the purpose of killing the fish.

THE MANNA-TREE.

During the intense heat of the month of August, when the sap is abundant, this tree furnishes a nutritive substance of slightly bitter taste. It is a natural secretion of the plant, and has procured for it the name which recalls the miraculous food of the Israelites in the desert. Manna is a liquid substance as clear as water, and flows from the tree, if about the middle of August an incision is made in the bark. Generally the first cut is made near the foot of the tree; and each day a new incision is added two inches above the last and so on up to the lower branches. These cuts, generally made with a pruning-knife or carpenter's chisel, are usually two inches in length and half an inch deep.

At first the sap flows abundantly, like a liquid stream; at the end of a month it becomes thicker and flows less freely. The rainy season interrupts it altogether, and towards the end of September the heat of the day is no longer sufficiently powerful to make the sap rise, which gradually recedes to the lowest parts of the tree.

The manna gradually loses its slightly bitter taste which it had when taken from the tree; the

MANNA TREE

watery particles evaporate, and its taste becomes even insipid and not at all appetizing.

This tree (*Fraximus ornus*) belongs to the same genus as the common ash, and is a native of Sicily and the south of Italy. Its normal height is 20 feet; at first view it might be taken for a young elm, but the appearance of the leaves clearly mark it as a different species. There are three varieties known; the leaves of the first are long and straight like peach leaves; those of the second resemble the leaves of the rose, while on the third variety they partake of the appearance of both the former.

The manna of Calabria is highly esteemed, and the most renowned kind grown there is that from the gardens of Œnotria. A popular tradition exists in the district to the effect that the kings of Naples having intended to enclose these gardens in order to raise a tax from the cultivation of manna, the latter suddenly ceased to flow, as if the trees had been struck with sterility, and they remained in this condition until the unjust impost had been removed.

It need hardly be added that the manna of the Israelites was in no way connected with the ash-tree of our day. What is now called manna by the Arab, is a gum from the tarfa or tamarisk shrub (*Tamarix gallica*), and the real manna of the Israelites has never yet been ascertained.

It is, however, by no means necessary for us to go to far distant lands in order to see what marvels nature displays by means of the simple sap which restlessly moves, like the blood of man, through every vein and artery of plants. In our own country, and strangely enough, in its most northern parts, we derive from this sap one of the most valuable products of the whole vegetable kingdom. Our sugar maple (*Acer saccharinum*), when tapped at the proper season, pours forth its liquid sweetness at the rate of a bucketful a day. This fluid, evaporated by gentle heat, yields a brown, luscious syrup, which is afterwards converted into diminutive sugar-cakes, and enters largely into home-consumption as maple-sugar throughout the whole northern portion of our Union.

THE TRAVELLER'S TREE—(*Urania Speciosa*).

This remarkable tree is found only on the island of Madagascar; it belongs to the Musaceæ and to the same family of which the banana, the plantains and the brilliant flowered strelitsias are members. Unlike other palm-trees, they grow better in the interior than upon the sea-shore; and their appearance produces an agreeable diversity among the bamboos, with their feathery-tufted clusters.

Travellers are unanimous in their grateful admiration for this tree, which, hence, has obtained the pet name of the *Traveller's Tree.* We are told that it grows principally in regions where there is no water, and that it has the admirable property of secreting for travellers a limpid and refreshing supply of water. Its large white leaves curve back towards the main trunk and thus form cavities in which the water is gathered and kept for the thirsty wayfarer. Some travellers have, however, failed to meet with this hospitable wonder

TRAVELLERS TREE.

of plants. Miss Ida Pfeiffer, who has been three times around the world, was unable to ascertain the facts; she even states that the natives of the country are not agreed on the subject, and assert that this tree only grows on moist soil. The Island of Madagascar is not yet sufficiently explored for botanists to be able to add their conclusive verdict upon its vegetable productions. On the other hand, there can be no doubt that a real weeping-tree (*Caesalpima pluviosa*) was seen some years ago in one of the Canary Islands, from the tufted foliage of which water fell like copious rain.

THE RAFFIA PALMS.

The raffia palms are more elegant; their long leaves are curved back as if to adorn with their graceful arabesques the summit of these lofty columns, which resemble the pillars of an edifice. Comparing this arrangement with the manner of building in the East, we are involuntarily led to believe that this vegetable architecture has furnished the original type of Byzantine architecture. The harmony of this natural temple seems to invite the mind to meditation and prayer, even more effectually than the Gothic stone arches which close the vault above us and prevent the aspirations of our hearts from rising heavenwards. In these palms every thing is of colossal size—a fact which we can, perhaps, best realize by seeing the spathe or envelope which protects the young flower, used as a cradle! Growing frequently to the size of a large cup nearly two yards long, it is used for various purposes.   In the accompanying illustration the thick, woody spathe of a large palm is thus made to serve as a bath for the children of native women.

SPATHE OF PALM TREE.

The Palm.

CHAPTER IV.

PALM-TREES.——THE DATE.

AFTER the trees of which we have spoken and which are mainly remarkable for their singularity, it is but right that we should begin our description of the vegetable world with the illustrious and ancient family of palms.

The " dynasty of palms," to use an expression of Linnæus, reigns over the tropical regions of our earth and occupies the highest rank among plants. This supremacy is due to them on account of their rich foliage, their beauty and elegance, and still more on account of the important services which they render the inhabitants of the Tropics. For these palms act-

ually supply the wants of their existence, furnishing
besides bread, oil, wine, also clothing and the common
tools and materials for building. Moreover, the palm
is a holy **tree** to many **races: to** the Mohammedans
the date palm is sacred as the **fruit** which Adam was
permitted to bring **with him out of** Eden—by the
Christian all palms are **revered as** having furnished
the leaves which were **strewn** in the path of the
Messiah.

In their form, aspect and structure, these plants
differ essentially from those of our country. A single
stem, straight and slender, rises to a height of 45, 60,
or even **75 feet**, perfectly bare, unbroken by a single
branch or leaf. At the top only an immense plume
of feathery leaves, growing in a bunch, forms, so to
speak, the capital of the vegetable column. This
tuft may be from nine to twelve feet long, and at the
roots of its long leaves appear the fruits of the palm-
tree. This short description applies especially to the
date palm, well known as the " prince of palms," and
hence as the prince of all plants.

Originally a native of Arabia and Northern Africa,
the date palm is pre-eminently the tree of the desert,
where it grows in nearly every oasis, and by its refreshing shade, its fruits, its milk, and its general usefulness, has won the affection of the natives and the
admiring sympathy of all travellers.

The date, says Mr. Ch. Martins, is the true friend
of the desert; there alone it ripens its fruit; and without it the Sahara would be uninhabitable. Arabic
poetry loves to praise it as a living being created by

God on the sixth day of creation, at the same time with man, to convey an idea of the conditions under which it thrives. The Saharans use a bold but expressive figure: "The king of the oasis," they say, "must plunge his feet in the water and his head in the fire of heaven." Science confirms this assertion in a manner, for it is proved that it requires 9.180° of accumulated heat, spread over eight months, to bring the fruits of the date to maturity.

"The climate of the Sahara fulfils these conditions," adds the great botanist. "The mean temperature required is from 68 to 73° according to locality. The heat commences in April and continues to October. During the summer the thermometer reaches 113° and even 125° in the shade. The winter is relatively cold. Dates can endure dry and short cold as low as 21° above zero and a heat of 122°. The radiating sand of the desert cools off more readily than the air, and preserves, at a certain depth, a degree of freshness which invigorates the roots of the trees. Rain is rare in the Sahara; it falls only in winter and woos the withered plants to a new life. Sometimes it rains in torrents; but these gusts are of short duration. At Tougourt and Ouargla, whole years pass without a drop of rain. Hence the very natural admiration of the Arab for this tree with its sweet fruits, which grows in the sand, fed by brackish waters that would be fatal to almost all other plants; which remains flourishing and green when all around is burnt up by the fierce rays of a pitiless sun; which resists the winds that may bend its pliant plume to the ground but cannot break its strong stipe, composed of interlaced fibres, nor tear it from the soil to which it clings by a thousand roots that strike deep and defy the tempest. It has well been said that a single tree has peopled the desert, and upon it alone is based a civilization, rudimentary compared to our own, but far advanced beyond the merely natural state. Its fruits are in demand throughout the whole world, suffice to procure all necessary imposts, and not only make the Arabs independent but affluent." In the 360 oases which belong to France, each date-tree is taxed from 20 to 60 centimes, according to the oasis, and their cultivation pays, for the mean produce of each tree is valued at about three francs.

We learn also from Martins that in order to obtain the milk of the date the Arabs of Tougourt employ the following means: They take off the crown of leaves, sparing only the lower ones. The section has the form of a cone, and into this a reed is then inserted, through which the liquid runs out into a vessel, which in its turn is emptied into another suspended from the leaves of the tree. The palm does not always die after the mutilation,—the terminal bud grows out again, and the tree gradually recovers and flourishes again. The process however cannot be repeated oftener than three times.

The tufts of palm-trees form a kind of vast parasol, under which the air can circulate; but the sun is unable to pierce the dense canopy of leaves. Shade, air and water are the three elements which permit the culti-

vation of many plants in palm gardens in spite of the
burning heat of summer.

An oasis of palms is a veritable paradise in the
burning waste of the desert. Such an oasis is graph-
ically described in the narrative of Mr. Martins, who
once accidentally discovered a clump of these marvellous trees during his passage across the Eastern Sahara. "The boundless desert," he says, "was stretching
out before me. The sun, high above the round horizon,—round as we see it on the ocean when out of sight
of land—seemed the only living thing in the midst of
death. All at once I perceived the summits of palms,
the trunks of which were not yet visible. I thought
it an illusion—a mirage. We drew nearer—the tufts
became more distinct, but the trunks could not yet be
seen. The caravan halts near a well. I hasten toward
the palms and find they are planted at the bottom of
a trough nearly 24 feet in depth. The sand had been
raised on all sides; a feeble palisade of palm leaves
helped to keep it up on one side, on the other sides
crystals of sulphate of lime of all sizes and shapes, arranged as we see them in collections of minerals, helped
to fix the shifting sand. At the bottom of the trough
the dates were planted irregularly; but this was not
the slender, elegant palm of the painter. These were
trees with short, thick trunks of cylindrical form; looking for all the world like the short, massive columns
of an Egyptian temple, or of a moorish mosque. Surface roots, joining the lower part of the trunk to the soil,
formed a pedestal for these columns, and the lofty tufts
on high resembled exactly the vast colonnades of an-

cient temples. In the evening, when penetrating under the sombre vaults of these palms, I could not resist a feeling of awe ; for these palms, majestic and immovable at the bottom of their crater of sand were a fit emblem of African civilization, unchanging amid the ever-changing outside world.

The family of palms is very numerous, and the different species which belong to it (450 have been counted) are of remarkable interest, both on account of their strange beauty and of the valuable services which they render to man. As the limits of this work do not permit us to examine all these treasures, we must content ourselves with a few that are most worthy of the interest and curiosity of our readers.

THE COCOA-NUT TREE.

Like the date-palm-tree, rises to a height of 90 feet, with a straight and smooth stem crowned with a capital of leaves in the shape of a plume—each leaf being about 18 feet long. It is met with throughout the whole Torrid Zone, but abounds chiefly near the sea-coast. All the wants of man, in his primitive condition, are supplied by a cocoa-nut palm—by its fruit, its seeds, its leaves, and the other parts of the plant. The following narrative, by M. Boniface Guizot, will give an excellent idea of the importance and the nature of its usefulness to man.

A traveller was journeying through those countries lying under a burning sun, where the freshness of shade is rare and the habitations of man are found only at considerable distances from each other. Sink-

ing with fatigue, the exhausted traveller beheld a hut
surrounded by trees with tall straight stems, surmounted by a bunch of great leaves, some standing
upright, others hanging down gracefully and presenting a beautiful and elegant appearance. Nothing
else near this cabin spoke of cultivation. Encouraged
by this sight, the traveller made a last struggle,
reached the hut, and was kindly received under the
hospitable roof. First his host offers him a slightly
acid drink, which quenched his thirst and refreshed
him. When the stranger had rested himself awhile,
the Indian invited him to partake of his repast,
and he served different dishes on a brown platter
shining brightly and highly polished; he offered him
also a wine possessing an extremely pleasant flavor.
Towards the end of the repast he brought a quantity
of excellent comfits, and invited him to try an excellent kind of brandy. The traveller in astonishment
asked the Indian how, in the midst of this desert, he
came by all these things.

"I get them from my cocoa-nut trees," answered
the Indian. "The water which I gave you on your
arrival was drawn from the fruit before it had become ripe, and sometimes the nut contains three or
four pounds of it. This palatable nut is the fruit at
its maturity; this milk, which you find so pleasant,
is drawn from the same ripe fruit; this delicate cabbage is made from the top leaves of the tree; but we
do not often indulge in this, as the tree, when its top
is thus cut off, dies soon after. This wine, which
pleases you so much, is also got from the cocoa. We

make an incision in the tender flower stalks and a white liquor flows forth, which we gather into vessels and which is known as palm wine. Exposed to the sun it becomes sour and turns into vinegar. When we distil it we obtain this excellent brandy, which you have tasted. The same juice has also furnished me with the sugar which I needed for preserving the nut. Finally, all these dishes and utensils which we are using on the table are made from the shells of the cocoa-nuts. This is not all—my house even I owe to these invaluable trees; their wood has enabled me to build my cabin; their leaves, dried and interwoven, make the roof; and these same leaves made into a parasol protect me from the sun when I walk out. These clothes which I wear are woven with the fibre-threads got from the leaves. Those sieves were ready made in the parts of the tree from which the leaves spring, and these mats come from the same source. These same leaves woven into a tissue make sails for our ships. The coarse hair which covers the nut is used for calking ships, as it lasts forever and swells when exposed to the water. Cables, ropes and twine are all made of the same material. Finally, the delicate oil with which many of these dishes were seasoned and which burns in my lamp, is obtained by pressing the freshly-gathered fruit."

The stranger listened with astonishment and wonder as the poor Indian showed him thus, that a single variety of palms furnished him not only all the necessaries but many of the luxuries of life. When the traveller was about to leave the cabin the host said to him:

"I wish to write to a friend in town, be good enough to carry my letter for me, I pray!"

"Most certainly; and is the cocoa to furnish you also your writing materials?"

"Certainly," answered the Indian; "from the saw-dust of the branches I have made this ink, and from the leaves this parchment, which formerly was exclusively used for public documents and records of important events."

THE LAQBY

At the time when the return of spring gives motion to the sluggish sap of the trees, a man mounts to the top of a date palm, climbing up the stem with no other assistance than what he obtains from his naked feet, and a cord passed round his waist and round the tree. He is armed with a very sharp hatchet. Arrived at the top, from which the rich plume of leaves rises proudly, he begins to hack away, without mercy, cutting off all the branches and leaving only four, which form a cross, and seem to point to the cardinal points of the compass. Over the neck of one of these he passes a slender cord, the ends of which reach to the ground, and between two of the remaining leaves he cuts deep into the poor wounded tree. The laqby cask is next broached. A small jar with a wide mouth is hoisted by means of the cord and is fixed to the mouth of the incision that has been made. Twelve hours afterward it can be taken down and replaced by another. It is full of a pale gray liquid, resembling weak barley-water. This is fresh laqby, a juice almost sickening from its excessive sweetness, but useful as a pleasant, weak laxative to be taken in the morning. A few hours later fermentation begins in the vessel, the liquid clears up and seems to boil; innumerable bubbles of air rise to the surface, forming a light foam, and if you taste the sparkling beverage now, you will not sigh for the vintage of champagne. Laqby, drunk in this condition, is harmless, cheering, without intoxication, and producing no evil effects; the fermentation renders it refreshing and takes away its laxative properties. But let it stand another half day and this liquor will become white and thick like milk, with a penetrating odor, and a slightly acrid taste, and in this state it intoxicates like brandy. The champagne has become a white beer of astonishing alcoholic strength. It is then that amateurs love it best. Many a good Mussulman and his scrupulous wife (who veils her face before a glass of wine), will drink in public, and without hesitation, a cup of laqby, which is only the " water of the palm."

When it has reached this stage the vessel must be emptied, for on the morrow the beverage would be found spoiled and full of small reddish insects. In fact it is the most perishable of all drinks, and has to be consumed under the very tree from which it is drawn. All attempts to regulate or arrest the fermentation have been fruitless. It preaches, like no other preacher, the poets' doctrine, *carpe diem.* In Tripoli (northern Africa), the **Arabs** drink their laqby daily, as they smoke their pipe contemplatively on the bank of a water-course.

ARECA PALM.

By the side of the champagne palm we must mention the slim areca palm, so highly esteemed by the Indians for its leaves and its fruits. In spite of its height, which often reaches 40 feet, the stem is exceedingly slender, and it is only by means of its deep roots that it can withstand the fierce winds of the tropics. Like all the trees of this class, the areca palm also is crowned with a magnificent plume of gigantic feathery leaves, some of which measure 15 feet in length ; if cut before they leave the massive bud in which they are at first carefully enclosed, they furnish the famous palm cabbage, a great favorite with Indians and Europeans alike.

A plantation of arecas is continually producing fruit, and often on the same tree three stages of ripeness may be observed in as many clusters of fruit. These fruits, when about the size of an egg, grow in groups, and assume, as they ripen, the color of an orange. They are sometimes gathered before being ripe, for the sake of their pulp, called *pinang*, which is then of an agreeable taste. But generally the growers wait till the usual six months bring perfect maturity, because the *pinang* is changed into a seed of the size of a nutmeg ; this nut is one of the three ingredients which make up the famous *betel*, so extensively chewed by the Indians, and which gives to their teeth a peculiar reddish black color, extremely repulsive to foreigners.

The betel is composed of small pieces of areca nut rolled up with a little fresh lime in the leaves of the betel pepper.

We are puzzled to know how these ingredients can be agreeable to the taste, yet it is certain that the practice of using the betel nut is of ancient origin among the East Indians, and at least as general there as the use of tobacco in Europe and America. Women even habitually use it; and the practice dates from such a remote time that there is no tradition among them that the teeth were ever allowed to retain their natural color. On the contrary, white teeth have ever been looked upon as extremely ugly, resembling dogs' teeth! All the effects of the betel nut are not, however, prejudicial. It strengthens the stomach and makes the breath very agreeable. But it destroys the enamel of the teeth and the teeth themselves—the lime having, probably, most to do with this effect.

The Indian betel must not be confounded with that used by the Turkish women; the latter has the advantage of the former in usefulness, and is said to possess none of its decided disadvantages.

The Indians always prepare their betel from the newly-gathered areca nut and the betel pepper. The color of betel is reddish, and gives the same color to the saliva; the latter has to be discarded till it loses its redness—a very inconvenient necessity, which however, does not hinder the Indian women from using it. The English often call the areca palm the betel-nut-tree—but the name does not at all belong to the tree, which is known to botanists only as the areca palm.

THE ELAÏS PALM.

Among the precious plants that grow in the forests of Africa, beyond Cape Verde, there is a palm, the leaves of which are spread out at the height of thirty feet from the ground, and which the natives call their *Friend*. Even those who have visited the splendid forests of the Tropics are struck with the beauty of this magnificent tree, the *Elaïs Guineensis*, which clothes all the slopes inclined towards the sea, and richly rewards the care bestowed upon it by the natives. And yet its beauty is by no means superior to its usefulness, as the exports from Liverpool to New York attest. And yet so far, of its many products, oil only has been an object of extended commerce and exportation. The natives, on the contrary, not only draw wine and oil from the noble tree, but they manufacture from it their fishing-lines, hats, baskets, wooden tools, and even timber for their houses. It is their companion and stay, charged by nature to subserve all their wants from day to day.

Formerly the manufacture of this palm oil was left entirely to the natives; but now it is carried on by foreigners in large farms among the forests of the coast. When the seeds are ripe they are gathered, cast into troughs and trampled under foot by negroes, who are provided with wooden sandals.

Palm oil is one of the most important products of the African coast. The elaïs does not grow under the same conditions as the sesamum which produces the gurgelly oil of commerce, for it is exclusively found in the tropical regions of Africa. It grows in large clusters in sheltered and fertile spots, and its magnificent appearance recalls that of the date palm of the Arabs. The oil is generally exported in its crude state and refined abroad, while at Marseilles it is made to produce soap also, and candles.

THE BOURBON PALM.

Linnæus gave to the palms the pompous title of the Princes of the Vegetable World. It may be truly said that they are the aristocracy of the world of plants, and on account of their beauty and majestic stature they are worthy of the name that he gave them.

The bourbon palm, and especially the red latanier, as it is called in Louisiana also, is one of the most beautiful of the family of palms. It came originally from the Southern provinces of China, and is spread over the whole of India. The flower is of a superb red color. The leaves are used by the natives to thatch their huts, and the fibres for the manufacture of light hats, as comfortable as our Panama hats, but very different in form and structure. This tree only flowers twice in a century.

The frontispiece of Hindoo manuscripts very frequently consists of a drawing representing the esteem in which palms are held in India: A man is seen reading, reposing under the shade of one of these trees. In fact, India is indebted to the palms not only

BOURBON PALM.

for the food of her children, but also for nearly all that is necessary for life. Three palms especially are of great service : the sago palm, the cocoa, and the date. The sago palm, when in bloom, gives to man a perfectly amazing amount of a farinaceous product. The cocoa, on the other hand, ministers quite alone to all his wants. Food, bread and wine, clothing, shelter, articles of daily use, all are provided by the cocoa. Nor is the date palm less valuable. We know what a wonderful source of food it is to the Africans. These three palms deserve the same consideration from the inhabitants of their respective countries which we give to wheat and the vine, and the natives are not ungrateful. In more than one ancient religion we find that these trees were the objects of adoration on the part of grateful nations.

The traveller in Palestine or Syria contemplates with a different interest the palms of the Holy Land. The date is the commonest tree in these regions. "Everywhere," says a recent traveller, "we see its round stipe balancing high in the air its ample clusters of fruit, and still higher above them its magnificent plume of leaves. Nothing is more beautiful than an avenue of these noble trees; and one can conceive the enthusiasm with which the prophets of the Bible and the poets of the East have celebrated it in their songs.

THE WAX PALM.

We cannot leave the chapter on palms without mentioning one which yields wax, the carnahuba

(*Corypha conifera*), to which Humboldt gave the name of the Tree of Life. It is one of these trees, says M. Denis in his work upon Brazil, which provide for all the wants of a village in the midst of a desert. Owing to the hardness of its wood and the manner in which its foliage is arranged, a commodious cabin can be constructed by the aid of two or three carnahuba, without employing any other material than a little mud to plaster the walls. The leaves are used in the manufacture of countless articles, mats, hats, ladders, baskets, and, in addition to all, serve as food for cattle. In times of great drought even the pith of the young wood can be given to cattle and they subsist upon it till better fodder can be had. Arrived at its full growth a nutritious starch is obtained from the tree, the far-famed farina of our day. Its fruit is pleasant, and would suit everybody, at least as long as it is not fully ripe. But the main production of the carnahuba, which gives it a special place in vegetable economy, is the wax which appears in the axes of the young leaves in the shape of a glutinous powder or larger fragments of irregular shape. This powder, obtained by the use of fire, assumes gradually the consistency and odor of wax. Small tapers are made of it in the countries where it is grown, and large quantities exported to Europe, to be mixed with tallow and manufactured into candles. This tree furnishes, moreover, beautifully mottled or clouded canes, which take a high polish and are eagerly sought after in commerce. Another wax palm (*Ceroxylon andicola*), growing on the highest table-lands of the Andes

PALMS OF THE SEYCHELLES.

and frequently reaching a height of 215 feet, changes its whole outer bark into wax, which the Indians scrape off, purify and change into candles or use as soap. It is in Havana that we admire to greatest advantage the beautiful family of palms. Palm avenues are often seen in the Island of Cuba planted in front of the white mansions that overlook the sugar-cane fields. Here are long avenues of palms, there of mangoes and orange-trees, and at the other end lie the gardens and vast plantations, where the negroes, men women and children, renew each day their labors.

In Cuba the air is not excessively hot, and yet perfectly transparent. Light clouds, says Mr. Dana, are floating at mid-height in the serene sky, the sun is brilliant, and the luxuriant flora of a perpetual summer covers the whole country. Everywhere rise these wondrous palms. Many of the other trees resemble ours, but these form the distinctive features of tropical climates. The royal palm, especially, is characteristic of the tropics—it cannot grow outside the narrow belt which encircles our globe, lying close to the Equator. It has no special beauty of its own, it gives no shade and bears no fruit that is useful to man, and yet, with all these disadvantages and drawbacks, it has the power to attract us irresistibly, and once seen it is never again forgotten.

Palm-trees are however not unknown in the United States; in Key West, for instance, cocoa-nut palms are seen overshadowing every thing, and presenting, in the young trees especially, such grace of form as vegetation shows nowhere else. Date palms also, bearing the dates of commerce, are quite numerous in Florida; and at Cape Sable there are groups of royal palms of matchless beauty, perhaps the first on earth through whose leaves the wires of the telegraph flashed their mysterious messages. Plants which furnish wax are found even in the United States. Such is, for instance, the *Myrica cerifera* of Louisiana, rising at times to a height of twelve feet, which was the first of its kind known in Europe, where the seed was imported and raised in hot-houses. A variety is found in the Middle States, Myrica, Pennsylvanica, Carolinensis, which does not grow above five feet high; the leaves are broader and stouter and the fruit is larger. Marshes and damp and sandy spots on the sea-shore are its favorite homes. A very fertile bush furnishes about seven pounds of berries, which produce nearly two pounds of wax. The latter is removed from the seeds by means of boiling water, in which the berries are violently shaken and bruised. Candles made from this vegetable wax perfume the room; they give a bright and clear light, especially if, as is usually the case here, a little tallow is added during the process. This wax myrtle, or candle-berry myrtle, enlivens the landscape by the brightness of its foliage, which is evergreen; it perfumes and purifies, by its balsamic exhalations, the insalubrious air of the marshes in the midst of which it flourishes.

We will close our remarks on palms by mentioning those of the Islands of the Seychelles, of which Pyrard de Laval, in his narrative of a voyage to the Maldives, says: "On the sea-coast is found a certain

ARBORESCENT FERNS.

nut, which the sea sometimes throws up, as large as
a man's head and looking very much like two melons
joined together. They call it *Tavarcarré*, and believe it to be the fruit of some trees growing beneath
the sea. The Portuguese call it the cocoa of the
Maldives. It has remarkable medicinal properties,
and is very dear. So precious was it, that when a
native wished to injure a neighbor, he accused him
of having found one of these nuts and concealed it
from the king, to whom it should be given up; and
when one became suddenly rich it was commonly
said that he had either found Tavarcarré or amber, as
if it were a priceless treasure."

The fruit of this palm was long known as *nux
medica*. The tree itself bears the name of Lodoicea.
Its huge fruit is often carried on the waves to considerable distances, and hence arose the idea that the nuts
were produced by submarine trees.

The realm of palm-trees is not without its mocking
imitations, in which nature occasionally seems to delight. A variety of ferns, called arborescent ferns,
and especially numerous in New Zealand, closely im-
itates the form and shape of palm-trees, and gives to
those distant landscapes an appearance utterly unlike
that of any other part of the world.

The Bamboo.

CHAPTER V.

BANANA, BAMBOO, BAOBAB.

THESE are perhaps the three strongest workmen employed in the vegetable world—they have withstood centuries, and no living thing rivals them in power.

Certain writers have tried to prove that the banana was the tree of life placed in the middle of paradise, the forbidden fruit, which, tempting the mother of the human race, "caused all our woes;" and besides, that it was of its leaves that Adam and Eve made themselves aprons when, after their sin, they dreaded to meet their Creator.

The inhabitants of America, Africa, and India,

and the natives of the Pacific islands fully appreciate
the immense value of this plant, which sustains a
large part of the races inhabiting tropical regions.

We have the banana, in Florida at least, as a plant
of our own also. Here, as everywhere, it is not a
tree, but annual in its growth, although the root is
perennial. In one year the banana grows from the
root to about twelve feet high, bears its one bunch of
fruit and dies. Other shoots are however coming up
in the mean time from the root; they in turn bear
their fruit, each after a year's growth, and this method of growing brings the plant into extensive and
beautiful groups. Every yard in Key West has its
banana patch, and the grand glossy leaves lend great
beauty to the humble cottage as well as to the imposing mansion.

For the plant sends up a single round and straight
stem of a yellowish green color, which terminates in
a fanlike expanse of large oval leaves, six feet long
and from eighteen to twenty inches in breadth. A
great strong midrib traverses the leaf, but the latter
is so tender that it is almost invariably torn into
shreds by the winds. The flower bud is purple, con-
trasting finely with the green of the leaves. It expands into a noble spike of flowers about four feet
high, rising from the centre of the leaves eight or
nine months after the planting of the vegetable.
The flowers are soon followed by the fruit, which is
eight inches long by one in diameter. These long
spikes of fruit sometimes weigh 70 pounds, and look
like a gigantic cluster of grapes formed of a large

number of fruits which frequently count as many as 150 or 160. When the tree is stripped of its fruits the stem also is cut down, which prevents the plant from drying up and causes the suckers at its base to grow up more rapidly, providing thus for another harvest six months afterward. The growing plant is aided from time to time by cultivating the soil around it, but this is all, and hence banana plantations usually placed near rivers are easily kept up with very little care. The dressing of bananas for the table is equally simple, as the fruit is cooked either in boiling water, on the oven or among hot ashes. The fibres of the stem are used to manufacture coarse shirts, and the green part is given as food to cattle. The inhabitants of the Moluccas subject the leaves to a certain process which enables them to convert them into a kind of linen.

Weight for weight the banana is inferior to wheat as nutritive food, but much more is produced on the same extent of ground. An acre of land planted in wheat would not yield sufficient to support two persons, whereas the same amount of land in the tropics, planted in bananas, would produce food enough for the support of fifty people! It has been calculated that a strip of land of two hundred square acres is capable of furnishing more than four thousand pounds of nutritive substance; from which it follows that the produce of this vegetable is to that of wheat sown upon an equal breadth of ground as 133 to 1, and to that of potatoes as 44 to 1.

In the abundant productions of the tropics we find

a striking comment upon human nature and the condition of its development. It proves that the progress of man is measured by the urgency and the contingencies of his necessities. The banana-tree feeds the inhabitants of the regions in which it grows without demanding labor—daily food is within their reach, sufficing for all their bodily wants without the necessity on their part of active exertions; consequently, they remain in a condition of comparative mental somnolence, and we find the character of their inert lives clearly written in their listless faces.

In Java there are bananas the appearance of which produces a deep and permanent impression upon the mind. M. de Molins thus describes his feelings upon arriving in the forests of that island:—

"After a journey of an hour and a half through the open country we found ourselves in the jungle. It was a confused mass of vegetation, in which, however, the wild banana, with its leaves a pale green on one side, and on the other spotted with red and brown, seemed to be the most prevalent tree. We steered our way through this sea of plants of all kinds, and admired in it above all the tree-ferns with their arborescent stems, and graceful and regular leaves—those marvellous ferns which vie equally with the flowers by their exquisite form, with the birds by their beautiful color, and with the trees by their imposing height.

"All at once the native who went with us as guide, and who was aware of the object of our expedition, stopped and called us: 'Look here!' 'Where?'

I asked. 'There,' he said, 'is the first of the giant trees, sir, the one you saw from town, sir.'

He pointed out to me a kind of tower adorned at the summit with branches and flowers, a structure such as no foreigner surely would ever have taken for a tree.

" This is only a small one," said the guide, " but in going higher up, you will find trees of larger growth."

In fact, although the specimen before our eyes seemed to be almost supernatural in its size, we saw as we came to the borders of the immense forest, that as we proceeded the trees became larger and larger still. One remarkable circumstance was that they were almost all diseased ; many of them were black at the top and stretched far into the air their huge, leafless arms. I was told that the sun was the cause of this, and that these vigorous trees could not endure the fierceness of its rays.

I am not able, now that I have no longer these giants of the forest before my eyes—to express the sense of awe excited in me by the sight of these colossi, veritable patriarchs of the forest, many of which, no doubt, had witnessed the earliest creations of nature, and belonged to epochs when the earth was still in its first vigorous youth, while now they surrounded me with their gigantic trunks and shaded me with the foliage of their enormous branches.

Humboldt· represents the bananas as everywhere found in company with palms. These trees, he says, are the ornaments of moist climates. Their fruits furnish the food of almost all the races that live in the Torrid Zone. As the farinaceous cereals have been an unfailing resource to the inhabitants of the North, the banana has never disappointed the nations that dwell near the equator. According to semitic traditions, this productive plant was first found upon the banks of the Euphrates; according to others it first grew in India, on the skirts of the Himalaya. Greek legends state that cereals first grew on the fields of Enna, in Sicily. But the fruits of Ceres, extended by cultivation to all the northern countries, present only monotonous fields, which add little to the picturesque charm of the landscape, while, on the other hand, the inhabitant of the tropics, who multiplies his banana-plantations, propagates one of the most beautiful and majestic forms of the vegetable kingdom.

BAMBOOS.

There is no tree known on earth which subserves so many purposes as the bamboo. The Indian obtains from it part of his food, many of his household utensils, and a wood at once lighter and capable of bearing greater strains than heavier timber of the same size. Besides, in expeditions in the tropics under the rays of a vertical sun, bamboo trunks have more than once been used as barrels, in which a water, much purer than could be preserved in vessels of any other kind, is kept fresh for the crew. Upon the west coast of South America, and in the large islands of Asia, bamboos furnish all the materials for the construction of houses at once pleasant, substantial, and

7

preferable to those of stone, which the frequently re-
curring earthquakes bring down upon the heads of the
lodgers.

An illustration of bamboos as they appear in the
tropics, heads the present chapter.

Leaving the immense size of these plants out of
consideration, **we would at the first** glance relegate
them either to the **class of grasses or of** reeds—their
appearance seeming to **indicate that** they belong to
former class, while the structure **of the** long hollow
stem, with its joints and sharp-pointed leaves, presents
all the characteristics of the latter. Botanists, however, have **at last** decided that bamboos are a tribe **of**
grasses.

But the name cannot alter the thing itself, and **it**
is not our purpose here to discuss the somewhat arbi-
trary classification of botanists. We prefer to con-
sider these plants simply as **we find them,** and to note
their distinctive characteristics without troubling ourselves about the Greek or Latin names which they
are made to bear.

These plants are found only in the Torrid Zone—
for the reason either that the heat of the tropics is
necessary to their development or that their cultivation has never yet been attempted in temperate climates under favorable circumstances. Of the 170
species discovered by modern travellers, five or six are
specially prominent.

The loftiest of the bamboos is the *Sammot.* In
the tracts where it grows in the greatest perfection it
sometimes rises to the height of 100 feet, with a stem

only 18 inches in diameter at the base. The wood itself is not more than an inch in thickness. The fact that the bamboo is hollow has made it eminently useful for a variety of purposes—it serves as a measure for liquids, and if fitted with a lid and a bottom, trunks and barrels are made of it. Small boats even are made of the largest trunks by strengthening them with strips of other wood where needed.

After the sammot, the next largest of the bamboo species is the *Illy*, which usually reaches a height of from 60 to 70 feet. It is used for the same purposes as the *Sammot*, and, like it, prefers a moist, rich soil.

The third variety prevails throughout Southern Asia, both on the continent and in the larger islands. It rises to the height of 50 feet. It is employed for the same purposes to which the other two varieties are applied, but is much more useful than either of these. For example, the young sprouts, of the stem and of the root, of the *Telin*—for such is the name given to this bamboo — are excellent food and are eaten as we eat asparagus, either prepared with vinegar and sauces or with other viands. European colonists are as fond of these shoots as the natives themselves. The wood of the *Telin* unites strength and lightness in a much more extraordinary degree than any other wood, and cut into thin planks or split into laths it is admirably suited for house-building in the tropics.

A still smaller species of the bamboo, which is not applied to so many purposes in domestic economy,

industry and agriculture, is the *Ampel*, which, however, furnishes levers, carts, ladders, and many similar
objects.   The Indians, when employed upon lofty
palm-trees collecting the palm-wine at a height of a 100
feet above the ground, are not afraid of going from
one tree to another means of a simple bridge made
of ampel-wood.   The airy bridge consists of a single
long stem of this tree and another lighter one serves
as a hand rail.   The young shoots, like those of the
telin, are used for food.   It is in this class of plants that
we meet with the iron-wood—as it is called in India
—which gives out sparks under the blows of a hatchet.
Its hardness is unequalled, and yet it can be split up
into the finest wands and in this form is much more
suitable for delicate basket-work than the osier.   Even
cloth of a certain kind is made from this bamboo.

We have still to mention the *Tcho* of the Chinese,
which furnishes them a solid paper, and is used in
manufacturing their large parasols.   Painters often
use it as canvas.   There is also the *Teba*, from which
hedges are made and the *Arundo scriptoria* of Linnæus, so called, because the Indian authors obtain their
pens from it.

These latter species prefer a dry, light soil, and
are equally acclimatized.   The sweet interior of their
young branches is a nourishing food, made use of by
man and also by herbivorous animals.   There is a
correspondence between the course of the moon and
the vegetation of these plants from which has arisen
the superstition that this satellite regulates their growth
by its influence, a superstition confined by no means

THE WONDERS OF VEGETATION.

to the Chinese, but quite common also among our negroes. The young shoots, which grow in bunches at the roots of the bamboos—the product of the underground germ—grow with such amazing rapidity that they may be literally said to be seen growing. In one day they obtain the height of several feet, and with the microscope, their development can be easily watched. But the most remarkable feature about the bamboo is their blossoming. With all this marvellous rapidity of growth they bloom only twice in a century, the flower appearing at the end of fifty years. Like other grasses, they die after having borne seed.

THE BAOBAB.

This plant of monstrous size, the most colossal and the most ancient vegetable monument on earth, has round, woolly leaves, which consist of from three to seven leaflets radiating from a common centre and giving them somewhat the appearance of a hand; and magnificent white flowers. It is an enormous tree, holding among plants the place that the elephant holds among animals—a hoary witness of the last changes which the earth has undergone and of deluges that have buried beneath their waves the productions of early ages. Several baobabs that have been measured were found to be from 70 to 77 feet in circumference. From its branches hang at times colossal nests, three feet in length, and resembling large oval baskets open at the bottom and looking from a distance like so many signal flags. The birds that build them are nearly the size of ostriches and thus correspond well with the giant tree that affords them shelter. The height of the baobab is, however, not in proportion to its circumference, as may be seen by our illustration.

It would take fifteen men with their arms extended to embrace the trunk of one of these great trees, which, in the countries through which the Senegal flows, are venerated as sacred **monuments.** Enormous branches are given off from **the central** stem a **few** feet above the ground and spread **out** horizontally, giving the tree a diameter of over **100 feet.** "Each of these branches," says M. Danton, "would be a monstrous tree elsewhere; and, taken together, **they** seem to make up a forest rather than a tree."

It **is** only at the age of 800 years that the baobabs attain their full size and then **cease** to grow.

The fruit of this **tree** is oblong; the color of the **shell** passes in ripening from green **to** yellow and brown. The fruit has been named "monkey bread." It contains a spongy substance, paler than chocolate and filled with abundant juice.

The bark is ashy-gray in color, and almost an inch in thickness. The negroes of the Senegal grind it down to powder, and in this state they use it to season their food and to maintain a moderately free perspiration, which enables them the more easily to withstand the excessive heat. It serves also as an antidote for certain fevers.

In Abyssinia bees choose baobab-trees for their hives, and their honey derives from the tree a perfume

THE BAOBAB

and a taste which make it to be much sought after
by the natives. Like the bees, poets and musicians
also are entombed by many African tribes in the
trunks of baobabs. In the eyes of Africans, however, these are not tombs of honor; and the reason
why they give their poets and musicians this strange
place of sepulchre is the belief that their gifted
brethren are in communication with spirits. They
have a superstitious horror of their remains, and will
not bury them in the earth that brings forth their
food, nor in the channels of rivers. It is difficult to
form an idea of the space which these trunks enclose;
some of them could hold 240 men. Besides using
them as places of sepulchre, the natives employ them
for other purposes; they sometimes encamp within
them, and at other times use them as stables.

Adanson has calculated the age of these trees by
the depth of certain notches made upon them by
sailors of the fifteenth century, who cut their names
in the bark in letters of considerable size; he examined the new layers of wood which had covered
these notches, and compared their thickness with that
of trunks of the same kind, the age of which was
known. "He has found," says Humboldt, "for a diameter of about 30 feet, an age of 5,150 years." He has,
however, had the prudence to add these words:
"The calculation of the age of each layer cannot be regarded as mathematically exact." In the village of
Grand Galarques, situated also in Senegambia, the negroes have ornamented the hollow of a baobab with
carvings cut in the wood. The interior space serves

as an assembly hall, where the affairs of the tribe are discussed. This hall recalls the "cavern" (specus) formed in the hollows of a palm-tree in Lycia, in which a consular personage, Licinius Mucianus, used to entertain nineteen friends at dinner. Pliny describes another cavity of the same kind as being eighty Roman feet in width.

The calculations of Adanson and of Perrottet, from which it would appear that there are baobabs in the world from 5,000 to 6,000 years old, would make these plants the contemporaries of the builders of the pyramids, or even of earlier mythical personages.

These immense trunks are crowned with a vast number of large, almost horizontal branches, and on this account they appear, when seen from a distance, like gigantic parasols; as the lower branches nearly reach down to the ground, they give to the whole form of the tree the appearance of a perfect hemisphere 100 feet in height and 250 feet in circumference.

The great dryness and intense heat of the tropical climate produce upon these trees the same effect which cold has upon northern plants; they lose their leaves, and only resume their foliage during the rainy season, which lasts from December to June.

Besides the uses which the negroes of Senegambia make of the fruit of the baobab, they are also careful to dry the leaves, which appear at this season, and to reduce them to powder, to which, as has been stated, they ascribe medicinal properties. It cures dysentery and the inflammatory fevers to which Europeans living in Senegal are frequently exposed.

The baobab surpasses all known trees in size, and even forms an exception to the general rule in vegetation in Australia. It is hardly ever found beyond a hundred miles from the coast, and it occurs most frequently on the river Glenelg as far as the western borders of Arnheim's Land. It prefers level sandy tracts; in stony and less fertile soil it rises to no great height, but still attains a colossal breadth, throwing out branches of extraordinary thickness. The fruit of the Australian baobab is much smaller than that of the African variety, in which an important trade is carried on in Senegal. But the fruit of the former is as highly prized by the Australians as the latter by the negroes of Senegambia. The tart pulp of this fruit is called by the German settlers on the Orange River, *Cream of Tartar,* and by the English colonists *Monkey bread.* The baobab of Australia is not considered as a curiosity only, but as a tree bearing a sort of providential food, which is obtained at once in a solid and liquid form, and a most abundant ministrant to human wants in that arid and burning climate.

The Cedars on Atlas Mountains.

CHAPTER VI.

CEDARS OF LEBANON AND OF AFRICA.

THE traveller who ascends the ancient mountains of Lebanon is overcome with awe when, having arrived at the lofty plateaux that crown them, he sees that the heavens are still shut out from his gaze by the green veil stretched above his head by the broad branches of the cedars. Calm and silent witnesses of revolutions that have altered the face of the world, they have beheld unmoved the terrors of man in the fearful days when the waters covered the earth. The strong men of the early ages of the world reposed under their shade, tribes set up their tents there, and patriarchal families rested there in their wanderings. As we approach them, we feel as if we were unworthy to touch them, so great in comparison with our little lives are the associations that crowd around these venerable giants.

" These trees are the most celebrated natural monuments in the world," says Lamartine, who visited them in 1833. " They have been alike consecrated by religion, poetry and history. The Scriptures celebrate them in many a passage, and they supplied the images which the poets delighted to use. Solomon wished to employ them in the building of the Temple, no doubt because of the magnificence and the sacred character of these trees even at this early epoch." The Arabs of all sects have a traditional veneration for these trees. They attribute to them not only a vegetative force, which enables them to live forever, but also a soul which imparts to them the power to manifest signs of consciousness and an understanding similar to the instinct of animals and the intelligence of man. They have a premonition of the seasons ; they move their huge branches like limbs—stretch them out and draw them in, raise them toward heaven or bend them toward the earth. In the Arab mind they are divine beings in the form of trees. They grow nowhere else but on the table-lands of the Lebanon, taking root high above the region where all other great plants cease to thrive.

The number of these trees diminishes in each succeeding age. In 1550, Bellon counted thirty of them ; in 1600, there were only twenty-four ; in 1650, twenty-two ; in 1700, sixteen ; in 1800, seven. These seven giant trees are, perhaps, the only living witnesses to-day of Biblical times.

Mount Lebanon separates the Holy Land from Syria, above whose loftiest mountains it towers. The range has the form of a horse-shoe, and measures not less than three thousand miles in length. To the south is Palestine, to the north Armenia, to the east Arabia, to the west the Syrian Sea. From Tripoli to Damascus, the slopes of the Lebanon are not far from the sea; at certain points they even touch the shore. The eastern part is known among the Greeks as the Anti-Lebanon.

The mountains rise the one above the other and present four different zones. According to travellers, the soil of the first zone produces grain crops and is rich in fruit-trees. The second zone is simply a belt of naked and sterile rocks. The third, in spite of its elevation, is covered with evergreens; and the softness of its temperature, its gardens, its orchards, filled with the finest fruit in all Syria, and the brooks which water it, make this a kind of earthly paradise. The fourth zone is in the clouds; and the perpetual snow, with which it is covered, has given the name of *Leban* (white) to these mountains. It is on one of the summits of this fourth zone that are still to be seen the cedars of Scripture.

"What prayers have ascended from beneath these branches!" exclaims the poet; "and where is there on earth a more beautiful temple than this one, so near to heaven itself? What daïs more majestic and more beautiful than this last plateau of the Lebanon!

What more productive of elevated thoughts than the cedar, the dome of which has sheltered and still shelters so many human generations, each one of which calls upon the name of God in a different tongue; but recognizes Him alike in all His works, and worships Him in the manifestation of his greatness in nature!"

The trees rise to the height of from sixty to one hundred feet. The largest of those that remain is thirteen feet in diameter and covers a circumference of one hundred and twenty feet. The branches, of a clear green even during the part of the year when they are covered with snow, are flat, horizontal, and covered with a close foliage. For a long time the cedar was classed as belonging to the larch family, but it is now regarded as a group of the *Pinus* family.

The fruit, as large as that of the pine, is rounder, more compact and smoother.

In his narrative of his journey to the Eastern Sahara, Mr. Martins speaks with the greatest admiration of the superb cedars of that part of the world. " The most beautiful forests of cedars," he says, "ornament the crests and the gorges of Chellalah, near Batna ; few are seen in Djurjura and at Teniet-el-Had, south of Miliana. What a contrast between these beautiful forests and the sterile tracts that lead to where they grow! When young, the cedars of the Atlas are pyramidal in form ; but, when they have grown taller than their neighbors or the rock which shelters them, a tempest, a thunderbolt, or an insect that pierces the terminal sprouts deprive them of their pointed tops. The branches spread sideways, and form a perfect wilderness of verdure, concealing the sky from the eyes of the traveller, who passes on in darkness under these vaults impenetrable to the rays of the sun. From the height of a lofty mountain-top the sight is still more imposing. The level surfaces of the trees then look like a broad dark green, or almost blue meadow, sown with egg-shaped purple cones. The eye is lost in an abyss of green, at the bottom of which an invisible torrent brawls along. Often an isolated group attracts the attention; as we draw near, expecting to see a number of trees, we are astonished to find that we stand before a single tree, cut down in times of yore by the Romans or the first Arab conquerors. The tree has sent up new shoots, enormous branches have grown from the old stock, and each is a tree of full-grown size, while vast fans of verdure spread out on all sides from the mutilated trunk and cast their shadows far across the earth. Some of these cedars are still standing, though dead; the bark has fallen off, and they stretch their bleached bare arms in all directions. The cedars of Africa still await their painter. A Marilhat has worthily painted the cedars of Lebanon; but his successors are content to paint over and over again the portraits of a few oaks of home forests, which the connoisseurs recognize as old friends in every exhibition. Eminent artists spend their lives in reproducing the same forms, while the venerable cedars live and die unknown in the gorges of the Atlas, where their beauty is admired only by the occasional traveller that ventures into these mountains."

The Cactus—the Giant Taper.

CHAPTER VII.

'THE SCREW-PINES.

THE astonishing diversity of the productions of nature in different climates is so great, that even ex-
perienced travellers cannot restrain an exclamation of
wonder, when they pass from one part of the world to
another, and even from one side to the other of the
same continent.   This is especially the case with explorers of Senegal, when they have just crossed the
desolate wastes of the Sahara.   The richest vegetation
suddenly succeeds to the most complete sterility, and
tall, black Africans are met with in place of the
stunted Arabs.   The trees preserve a never failing
freshness—growing young again each season before

they have time to assume an aged look. They are seen inclining toward the waves of the ocean, as if they desired to drink the tepid and saline waters.

The strange plant which we present in the accompanying illustration, belongs to the family of screw-pines (*Pandanaceae*), of which Senegambia is the favorite country, but which is also found in Polynesia, in New Zealand, and in Guinea. M. de Folin, who has drawn it from nature, gives the following details respecting it, as he observed it in Prince's Island, thirty hours' sail from the Guinea coast and 1° 30 N. Lat:

" A stream that falls from the steep cliffs of the island, dashing its silvery waves from rock to rock, keeps up a constant moisture in a narrow valley, where the heat of the rays of the sun that beat all day upon the cliffs on each side, is reflected and concentrated. The warm moist atmosphere, due to these causes, maintains a most vigorous vegetation at the foot of the valley. The screw-pine grows at a spot where the gorge widens, and the torrent, spreading out into a limpid lake, pauses for a moment before flowing forth to fall into the sea.

This strange tree, with its slender supports, its bare branches, gracefully inclining toward the horizon, and spreading out its enormous fans and diadems of beautiful leaves, has the most airy appearance.

Masses of young shoots and of aquatic plants are grouped around each trunk and reflected in the water that furnishes the screw-pine its home and its support.

The weirdness of the strange scene is heightened

THE SCREW PINE

by the solitude that reigns all around in perfect si-
lence ; only now and then some aquatic animal utters
a low cry as it throws itself upon the shore, or a lonely
heron, perched upon a half-submerged rock, exults as
he swoops down furiously upon his prey.

Among the screw-pines we must notice one species much prized by the inhabitants of Oceanica, who
weave beautiful mats with its leaves. It is called
the sweet-scented Pandanus (*P. odoratissimus*), from
the circumstance that its flowers exhale an odor at
once sweet and strong, which perfumes the whole
neighborhood. Another screw-pine, more extraordinary still, if we are to believe De Candolle, has a
flower which in opening emits a flash of light accompanied by sound.

In Madagascar is found the *Pandanus muricatus ;*
but we look in vain in this island for the beautiful
trees which adorn the virgin forests of Sumatra, of
Borneo, or even America. Yet the useful screw-
pines overrun the low reaches of the coast. They
are of a singular form, full of grace, and yet mournful.
The trunk, covered with a smooth bark, divides at
the height of about six feet into three branches of equal
size. Each branch divided again into three others,
forms thus at the summit a crown of the finest foliage.
The entire height never exceeds thirty feet.

THE CACTUS.—THE GIANT CANDLE.

In America, from the Mississippi to the shores of
the Pacific, in the state of Sonora, and in Southern
California, the traveller meets with the gigantic candle-

tree (*Cereus giganteus*), a plant which is at once sim
ple and singular in form, and is called "the giant can
dle," because of its form and height.   It is the queen of
the cactus tribe, and towers with its straight stem above
the short and twisted varieties that belong to the same
family.

"In this country," says the traveller Möllhausen,
"animals and plants show to advantage, though the
same thing cannot be said of the human inhabitants.
The hideous Indians whom we met, dwelt near a de-
file called the Cactus Pass, because of the plants of
that name, that are found there in great numbers."
Among these the most remarkable is the *Cereus giganteus*.   This king of the cacti is known in California and in New Mexico as the *Petahoya*.   The
missionaries who more than a century ago reached the
Colorado and the Gila, speak of the fruit of the *Petahoya* upon which the natives subsisted and with
which they were as much delighted as in later days
the trappers.   This strange plant consists of nothing
but a few branches and still fewer leaves.   Its northern limit reaches to the banks of the Gila.   Savage
deserts and the most sterile tracts seem to be the
localities most favored by this plant, which finds
means of pushing its roots between stones and rocks,
where not an atom of soil is to be seen and where it
grows, nevertheless, to a surprising height.   The form
of these cacti varies with age.   At first twice as large
at the top as at the root, the plant, in proportion as it arrives at maturity, enlarges it diameter till it becomes
symmetrical and assumes the appearance of a straight

ASCLEPIAS GIGANTEA.

column for about twenty feet, to the place where the branches are produced. Here round branches go straight out from the trunk, but they gradually curve upwards, parallel to the trunk, and rise to the same height. It is at this stage that the curious plant, with its many upright branches, looks like a gigantic candelabrum, and deserves its name of giant candle.

At first sight one cannot conceive how these lofty stems, isolated, and clinging only to a point of rock, can withstand the tempest; but they owe their security to a series of ribs placed in the interior of the fleshy stem, from the top to the bottom, and which are as hard as the wood of the cactus. Both trunk and branches are regularly fluted throughout their entire length, and from this circumstance they bear a striking likeness to Corinthian columns. In May and June, the time of bloom, the upper end of the branches, and of the main stem, are covered with large white flowers, which are replaced in the two following months by savory fruits. This plant is one of the favorite articles of food used by the Indians, who also convert it into a sort of syrup. Upon the tree these oval and pear-shaped fruits grow close together; they are green, but at the top turn reddish. The pulp is crimson, and tastes like that of the fresh fig, but much drier. These cacti reach the height of 60 feet; when the plant dies, the flesh falls away, piece by piece, from the fibres of the stem, and for years afterwards holding on by the roots, these gigantic and bare skeletons are seen still clinging to the rock.

"It is to the New World," says Humboldt, "that the cactus-form exclusively belongs; they appear sometimes jointed, sometimes spherical, and sometimes like fluted columns, or organ tubes." This group forms the most striking contrast with the lily-tribe and the bananas. It belongs to that class of plants which Saint Pierre named the "vegetable springs of the desert." In the arid plains of South America, the animals, tormented with thirst, dig under the sand for the melo-cactus, the watery pith of which is protected by formidable thorns. The cacti, which take the form of pillars, reach a height of 27 or 30 feet. Divided into branches like candelabra, and often covered with lichens, they present an appearance like that of some of the euphorbias of Africa. These plants form vast oases in the midst of deserts bare of all vegetation.

The flowers of the night-blooming cacti have everywhere been regarded as symbolical. The *cereus* obtained its name from the torches with which Ceres is said to have searched for Proserpine. The superb cactus, which is called the torch-thistle in Mexico, is called the steppe-light in Russia. Our own Indians, and those of South America, seem to have observed the phenomena of sleeping and night-blooming plants, and it has been thought that they had to some extent anticipated the famous floral clock of Linnæus.

ASCLEPIAS GIGANTEA.

In the aspect of its trees, Eastern Africa presents to us forms not less strange than the names which they bear. South of the Strait of Bab-el-Mandeb (Strait of Tears), near the Guhet-el-Kherab (Basin of Untruth), which is a small bay of that part of the Arabian Gulf called Bahr-el-Bonatien (Sea of the Two Sisters), there stands the little town of Tanjourra. It is here, especially, that the *Asclepias gigantea* is found growing; a prickly acacia covered with a number of exuberant lianes. The small antelope, as well as aquatic fowls and the water-hen, haunt the shady woods formed by these beautiful trees; and this calm and enchanting scene would leave no unpleasant impression on your mind, if Tanjourra were not the centre of an abominable slave-trade.

THE CORK OAK.

We will conclude this chapter with a useful plant better known by its peculiar product than in itself. The description of its bark will lead us to consider the general structure of all trees.

A section of a full-grown tree presents three fundamental concentric subdivisions. First, the medullary canal, containing the pith or medulla. Second, the woody substance surrounding the pith. Third, an outer envelope—the bark. In the bark itself there are again three different substances placed in juxtaposition; the liber, consisting of thin leaflets, the parenchyme or cellular system, through which the sap circulates, and the epidermis or outer skin. This is the general structure of all trees. In the tree which produces cork, the parenchyme or middle division of the bark is the portion which furnishes that substance.

It is only after the cork tree is fifteen years of age that it has a parenchyma sufficiently developed to serve for this purpose. From this time onward to its last years we can strip the tree of its bark every eight or ten years, and each barking will produce 90 or 110 pounds of cork. In Catalonia, the true home of the cork-tree, or the cork oak as it is also called, a sufficient quantity of cork is reaped every year for the manufacture of 500,000,000 of corks, which are put up in bales of 30,000 each.

The manner in which the cork is gathered is thus: two incisions are made in the bark round the tree, and then two perpendicular incisions, taking care not to reach the liber—the innermost layer of the bark. Through one of the horizontal cuts a thin sharp blade is introduced and a square piece of the bark carefully removed. Other incisions are made and other squares of cork removed from the tree until it has been completely stripped. A liquid resembling melted wax flows in between the liber and the parenchyma and facilitates the operation. After being stripped, the cork oak is soon covered again with a viscous matter which escapes from tiny openings in the liber, and which spreads over the surface, hardens, and forms the basis of a new bark. But there must be an interval of about ten years before the tree can be stripped again.

This tree belongs specially to warm climates, and Algeria possesses whole forests which are now being worked by French colonists.

THE WEEPING TREE

A savage shooting poisoned Arrows (p. 139).

CHAPTER VIII.

MILKY SECRETIONS.

THOSE milk-trees which we have described in the first pages of this book, are not the only ones which are remarkable for an abundance of milky sap. Others also, serving purposes of another nature or even of pernicious and fatal character, deserve to be classed among the vegetables worthy of our attention. The families of plants which are most remarkable for their abundant sap are the *Euphorbiaceæ*, the *Apocyneæ* and the *Urticeæ*—differing from each other in their anatomical structure. They have in their bark and sometimes in the pith of their stems, a number of long tubes, more or less inosculated and flexible, which are so much like the veins of animals that they have misled many a naturalist and justified the comparison of vegetable sap with animal blood. Yet it seems that the term of " vital fluid," as applied to the latter, is inappropriate, and that of milky sap is more justifiable.

Certain trees which contain milky sap in great quantity have been called the serpents of the vegetable kingdom; and the most striking feature of the resemblance is in the organ, by the help of which both the plant and the animal emit poison. It is well known that with many serpents the poison is held in two long teeth of the upper jaw, which are traversed throughout their whole length by a narrow canal. At the root of these teeth is a gland that secretes the poison, and can be compressed by the pressure of the teeth like a sponge. At the moment when the animal bites, the poison is thrown into the medullary canal of the tooth and through a small opening into the wound. In poisonous plants we observe a similar arrangement in the bristles of the leaves—we can easily see this by examining the leaves of a nettle. The poison of the common nettle is as little dangerous as that of many snakes, but it becomes deadly as we approach the equator, the heat of the tropical sun seeming to intensify the venom both of the plant and the snake.

The three great families which are distinguished for the abundance and the value of their milky juices resemble each other in the nature of that liquid; and hence we shall here mention only the most remarkable species. Foremost among these stands a vegeta-

GUTTA-PERCHA TREE

ble product the employment of which for various uses
has been wonderfully extended in our days—the In-
dia-rubber or caoutchouc.

This gum can be obtained from a great number
of trees ; those which produce it in greatest abundance
being *Hevea-guyanensis*, the *Siphonia cahuchu* and
the *Jatropha elastica*. In the Antilles it is extract-
ed from the purple Euphorbia and the elastic Urceale,
the product of which is esteemed by some superior to
that of the Hevea. In spite of this great number of
caoutchouc-plants one would almost fear that the immense quantities of caoutchouc brought into all the
markets of the world, would soon transform the forests in which they grow into wastes of dead trees, as
has happened in North Carolina, where the larches
and pines, which have been tapped for their turpen-
tine, covers vast territories with dead wood, looking
like forests of bare masts.

The infinitely multiplied uses to which caoutchouc
is applied in these days are truly remarkable. In
England and America it is used to an enormous extent. In 1820, 52,000 pounds were imported into
Great Britain; in 1833, 178,676 pounds; and at the
present day a much larger quantity is imported. The
United States consumes more than twice this quantity.

This increase is of course much more noticeable
since the invention of vulcanite or vulcanized caoutchouc. The vulcanization is a chemical process, the
effect of which is to remove entirely the elasticity of
the material and to give it the various qualities possessed by wood, tortoise-shell, ivory or whalebone, and

9

to render it capable of enduring unharmed a high degree of heat as well as of cold, and of resisting moisture as well as the contact with acids. This effect is
obtained by combining it with sulphur either directly
or by means of bisulphide of carbon. Every one
knows the quantity and the diversity of objects that
are made from this light and yet hard vulcanite, from articles of jewelry and ornament to scientific instruments
and the tools used in general industry. In fact In-
dia-rubber and its more recent brother gutta-percha,
assumes a greater number of transformations than the
magic wand of the most potent fairy ever brought
about in Arabian tales. They run through the entire
list of useful and ornamental articles, from the breastpin
tipped with gold to the life-boat in the surges of the
ocean.

It was in 1736 that Condamine sent the first relia-
ble account of the new substance to the French Acad-
emy, describing it as the inspissated juice of a tree
called by the natives *Hevee.* In 1757 Fremeau found
the same tree in Cayenne, and it is now known to be
the produce of many trees growing in South America
and the East Indies. The most important of these is
one of the spurge tribe, *the Siphonia elastica,* found
in the dense forests on the banks of the Amazon, and
yielding the caoutchouc of Para; the Pernambuco
caoutchouc is furnished by the *Hancomia speciosa,*
found about Pernambuco and Bahia; the *Ticus elastica,* or snake-tree, with a wood so light and porous as to
be fit only for fuel or charcoal, produces an abundant
supply of milk, which the natives use for lining the

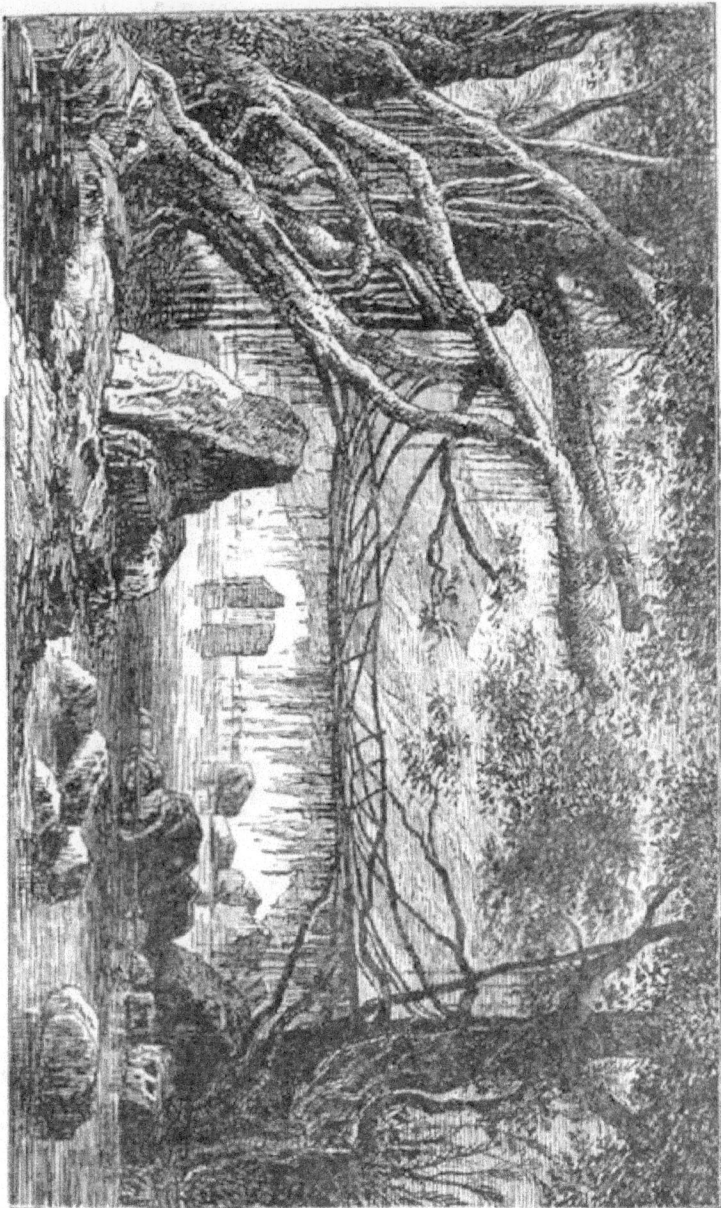

CAOUTCHOUC TREE.

inside of their water-pots, and making the caoutchouc itself into candles. A kind of junglevine (*urceola elastica*), of the Prince of Wales Island, is the main representative of this class in that remote portion of our globe. Caoutchouc is obtained in the following manner: with a sharp instrument straight and sloping incisions are made one above the other, the first about a man's height from the ground, which penetrate beneath the bark. At the foot of the incision a vessel made of clay, and holding about a tumblerful, is placed to receive all the sap; these bowls are filled in about three hours, if the tree is good, and from this the milk is poured into a calabash at the foot of the tree. The sap is liquid, and generally white at the time of extraction; the brown color with which we are familiar is imparted to it by foreign matter, which is mixed up with it, and it is still further darkened by the fires of Urucari nuts, which yield a thick oily smoke and are said to be of great value in the process. The Indians have clay moulds of bottles, animals, etc., which they dip into the milk and hold over the smoke till dry, repeating this until the rubber is of sufficient thickness, when they take it off the mould and the native manufacture is at an end. The caoutchouc is suspended in the albumen of the sap, like cream in milk. In order to separate the caoutchouc from the other matter, the whole is put in three or four times its bulk of water, and the valuable material rising to the surface is removed on the following day.

All the countries that produce caoutchouc are within the Torrid Zone; these are chiefly South

America, the East Indies, and certain parts of Africa.
On this subject, Humboldt states that the number of
lactiferous plants **increase as we near the equator.**
The heat of the tropics seems to exercise a great influence on **the** production of caoutchouc; for it has
been observed that the plants which produce it under
the tropics, when cultivated **in** northern climates,
yield a substance which **resembles** the glue of the
mistletoe, **but is** useless as **caoutchouc.**

EUPHORBIACEÆ.—MANIOC.—MANCHINEEL.

Although certain euphorbiaceæ yield caoutchouc,
yet others of that same family of which **we are**
about to speak, give different products. The juice
of the *Taybaba dolce* (*Euphorbia balsamifera*), resembles **fresh** milk; and Leopold von Buch relates
that the natives make **a** jelly of it, which they esteem
very highly. But all the euphorbiaceæ are not equally
harmless; some contain a **virulent poison,** and what
is very remarkable, **some of** them **yield** at the same
time a deadly **poison, and a thoroughly** wholesome
article of food.

The culture **of** tapioca or cassava is in **Central**
America what the culture of the cereals is in Europe.
There is, however, a great difference in the varieties
of this plant. The sweet cassava is eaten as whole-
some food, **while the** bitter cassava contains a deadly
poison. Let us follow the **natives into** their camp for
a little, **with** Schleiden, the **author** of "The Plant
and Its Life," and see what uses they make of this
remarkable plant.

In the midst of a dense forest in Guiana, the chief of the tribe, after having stretched his hammock between two great magnolias, **rests** under the shade of the large leaves **of** a banana-tree. He smokes indo-**lently, and** watches the movements going on around **him.** Meanwhile his wife crushes the manioc roots **she** has painfully gathered, in the hollow of a tree, by means of a wooden pestle; wraps the pulp in a net made of the leaves of a large lily, and suspends it upon a fork, tying a heavy stone to the lower part, so as to compress the contents and to squeeze out all the juice of the manioc. This juice, as it drops, is received in a calabash; and a little boy squatting by its side, steeps his father's arrows in the deadly liquid as it drops down; while the mother makes a fire to roast the strained porridge, and thus to rid it entirely of its volatile poison. After this, being ground to powder between two stones, the cassava flour is ready for domestic use.

Meanwhile the boy also has finished his dangerous **task. The** juice has deposited a delicate white starch, **which he** separates from the liquid, and which, after **having been washed once** more in fresh water turns **out to be tapioca! In** this or a similar manner the manioc and **tapioca of the** tropics are prepared **ev-**erywhere.

The savage, having satisfied his hunger, saunters about in search of a new resting-place; but woe to him! he has chanced to encamp under a redoubtable manchineel tree; rain suddenly falls, dropping from its **leaves upon** him, and the unfortunate man awakens suddenly, with pain racking his limbs, and blisters and ulcers covering his whole body. If he escapes with his life, it will be to cherish during the remainder of his days a wholesome fear of the danger of the poisonous properties of the euphorbiaceæ.

Everywhere the manchineel enjoys the unenviable reputation of being a most dangerous tree, in the shade of which it is imprudent to repose, since, as the poet says, "pleasure dwells there by the side of death." This evil reputation has its origin in the poisonous qualities of the sap and the fruit of a tree of the same kind, found in Africa—the arborescent euphorbia. Like the former, this tree has a magnificent though even more peculiar appearance. The thickness of its branches and foliage, which wholly exclude the sun, seem to invite the weary traveller to repose. The negroes have a way of taking advantage of the delightfully cool shade and at the same time avoiding the danger from the poisonous droppings of the tree. They erect a thatched roof below the lowest branches and repose in peace.

M. Trémaux, in a narrative of his excursion to the Soudan, has an interesting passage respecting these arborescent euphorbia.

"While taking a view of Cacane," he says, " I asked one of the negroes who stood near me, to go and seat himself under a great euphorbia which stood in the foreground. At first he hesitated; then after a little he decided to yield; but not without raising his eyes many times in apprehension towards the branches of the tree. I was about to climb upon a rock in order to break off a branch which I brought home with me to France, but the negro seeing me approach, fled in terror from the shade of the deadly tree, gesticulating wildly and shouting words in a language I could not comprehend. His signs, however, and a few Arab words uttered by one of the bystanders: 'Do you mean to die?' made me understand that in touching the tree I was running a serious danger. But the thing was done and the broken branch in my hand; immediately a milky liquid flowed forth —in much greater quantity than I could have imagined from what I knew of these plants in other countries—covering my clothes and penetrating even to my skin. The features and gestures of the negroes expressed their pity and their fear. They made me understand, that if the white juice touched one of the numerous wounds which I at that time had on my body, I should die; and that it was dangerous even to let it touch the skin.

"It is with this juice that they poison their weapons in order to make their wounds mortal; but they first thicken it, till it acquires the consistency of paste; then they dip in it the points or blades of the weapons they wish to poison."

Trees of this kind often reach twenty-four feet in diameter, and seventy feet in circumference. The greatest height of trees of this size is twenty-four feet. The trunk and the large branches are of hard wood; the smaller branches consist mostly of pith and parenchyme, sustained by a slender woody fibre.

TREACHEROUS PLANTS.

The remarkable quality which we have mentioned in speaking of the " Treacherous Plants," producing at one and the same time wholesome food and a terrible poison, is even more characteristic in a more striking degree of another class of plants. The milky juice of some of these is rich in caoutchouc; in others it appears in the form of sweet milk, wholesome and palatable, and in a third variety it assumes the form of a deadly poison. We have spoken already of milk-trees proper, of trees producing caoutchouc, and of arborescent euphorbias, but many of these plants are more deadly than any we have yet mentioned. The savages of South America poison their arrows with euphorbia-milk, and the natives of Ethiopia do the same at the Cape; they employ pieces of meat powdered with the pollen of *Hyananche globosa*, as an infallible means of killing hyenas.

One species of euphorbia described by Martins presents this remarkable peculiarity, that its milk, when it is drawn from the tree in dark warm summer nights, gives out a phosphorescent light.

The woorare, ourari, urali, etc., are nothing else than the *carare*. In past times this substance was believed to consist of a vegetable juice mixed with the blood of the viper, the poison of the rattlesnake, the saliva of serpents and other poisonous substances. These statements were shown to be false by Humboldt, Boussingault and other travellers, who have had an opportunity of studying the plants which produce it, the mode of extraction which the Indians follow, and their employment in the hands of those who make cruel use of it. It is a purely vegetable substance, produced by a liane belonging to the genus Strychnos, and abounding east of the mission of Esmaralda, on the left bank of the Orinoco, but growing also on the eastern slopes of the Cordilleras and in the forests upon the banks of the great equatorial rivers of South America. It is called the mavacure liana (*Strychnos toxifera*)

" When the bark of the mavacure is opened a yellowish liquid," says Humboldt, " continues to ooze out for several hours drop by drop. This filtered juice is the poisonous liquid ; but it has not acquired all its strength until it is concentrated by evaporation in a large clay pot placed over a fire. The Indian who filled the office of Poison Master, asked me from time to time to taste this poison liquid. It is by the bitterness of the taste that one judges whether the poison has been sufficiently concentrated. There is no danger in tasting curare, as it becomes fatal only by coming in direct contact with the blood."

Other travellers, like Scomburgk and Poeppig, have given us interesting descriptions of this preparation and of the deadly properties of the poison, which are so overwhelming that the Indians still use it in preference to the fire-arms of Europeans. The savage arms himself with a long and straight tube ; the points of his arrows, made of hard wood and a foot long, are dipped in the curare, while the other end is wrapped in a quantity of cotton, which makes it exactly fit the tube. With this terrible weapon he endeavors to surprise his enemy—perhaps some monster of the wood, who, having captured a deer, is tranquilly regaling himself upon the body. Not the slightest noise betrays the approach of the practised, cautious footstep; no eye beholds the long slender tube; and the winged messenger of death, propelled by the silent breath of the Indian, reaches sometimes after a flight of thirty paces, its unsuspecting victim with unerring certainty. However small the wound may be, the animal falls to the ground in awful convulsions and dies in a few minutes.

Schleiden states that a multitude of plants of the same family contain similar poisons. Here, however, the poisonous quality rests in their seeds, and this circumstance distinguishes them from those we have mentioned. In the form of strychnine and buncine, they present to us the two most violent of all vegetable poisons. The bean of S. Ignatius (*Ignatius amara*) growing in Manilla, and the nux vomica (*Strychnos nux vomica*) are found everywhere in the tropics. The natives of Madagascar have a custom recalling the ordeals of Europe in the middle ages, by which they make the guilt or innocence of a person depend upon the strength of his stomach. The man accused of a crime, is obliged, in the presence of the people and the priests, to swallow a Thangiu nut; if his stomach is strong enough to vomit up the terrible poison, he is acquitted; but if not, he is held guilty, and immediately made to undergo his punishment, for he dies on the instant.

POISON-TREES OF JAVA.—*The Upas-Tree.*

Many trees produce poison like the curare that grows on the Orinoco, and the woorare, which is found on the banks of the Amazon ; but the most terrible of all is the upas, which grows in several parts of East India, in Java, Borneo, Sumatra and the Celebes.

The Deadly Upas.

Rumph, who has given us a description of it, calls it the *arbor toxicari*. This tree has a thick stem and extended branches. Its bark is brown and knotty; its wood is hard, of a pale yellow color, and marked with black spots. Of all the different species of strychnos (from which we obtain strychnine), the upas and the nux vomica furnish the most violent poison. Astonishing facts, and still more astonishing fictions are told of this wonderful plant ; we will throw aside the fictions, and endeavor to find in

the facts enough to interest us. For recent travellers
have torn the veil of fables which has long surrounded and concealed the true nature of this remarkable tree, but enough is left to engage our attention.
We give first, under reserve, what is said of it by
Thunberg, a famous botanist of Upsaloa, Sweden.

"The upas-tree, an evergreen," he says, "is easily recognized at a great distance. The ground
around it is sterile, and looks as if it had been
burned. The sap is of a dark-brown color, and becomes liquid by heat, like other resins. Those
who gather it have to employ the greatest care;
covering the head, the hands, and the whole body,
to protect themselves from the poisonous emanations of the tree, and especially from the drops
which fall from it. They avoid even approaching
too near, and they provide themselves with bamboos tipped with steel heads, having a groove in
the middle. A score of these long spears are struck
into the tree, and the sap runs down the groove into the hollow bamboo, until it is stopped by the
first joint of the wood. The spears are left sticking
in the trunk for three or four hours, so that the sap
may fill up the space prepared for it, and have time
to harden, after which they are drawn out. The
part of the bamboo which contains the poison is
then broken off, and covered up with great care.
If kept for a year or two, the poison loses its vir-
ulence.

"The sap of the upas-tree produces spasms and
prostration. Persons passing beneath its branches,

UPAS TREE

bare-headed, lose their hair. A single drop falling
upon the skin produces inflammation. Birds can with
difficulty fly over the tree, and if they by any chance
alight on its branches, they fall dead. The soil
around is perfectly sterile to the distance of a stone's
throw. Any one wounded with a dart poisoned with
this resin is attacked by violent inflammation, followed
by convulsions, and dies in less than 15 minutes.
After death the skin is covered with dark spots, the
face becomes livid and swollen, and the whites of
the eyes turn yellow."

Foerset speaks thus of the effects of the juice of
the upas. "Being at Soura Charta," he says, "I
was present at the execution of three women. They
were conducted at 11 A. M., to the square opposite
the palace. The judge passed sentence upon them.
They were presented with the Koran, upon which
they had to swear that their sentence was just, and
this they did, placing one hand upon the book, the
other upon their breast, and raising their eyes to
heaven. Afterward the executioner proceeded to his
grim business in the following manner:

"Three stakes had been prepared, and to these
the convicts were bound. They remained in this position, saying their prayers, until the judge gave the
signal, when the executioner pricked each of them
in the breast with a lancet that had been dipped in
the resin of the upas. Instantly they were seized
with violent trembling, then with convulsions, and
in six minutes neither of the three survived. I saw
that their skin was marked with livid spots, their

10

faces were swollen, their color bluish, and their eyes yellow.

"I had occasion to witness another execution at Samarang, when seven Malays were put to death, and the effects of the poison were just the same."

The Dutch writer gives additional narratives, which we must, however, look upon as fabulous; but as in the foregoing, he deals with facts which are confirmed by other writers, and are explained by the known violence of the poison, we have mentioned his statements.

The forests of Java present little that is attractive to European explorers, and in passing through them, a feeling of fear is mingled with curiosity. "On all sides," says Schleiden, "palm-trees armed with thorns and long prickles; seeds with their edged leaves sharp as knives, repel with their dangerous weapons, all those who attempt to pass into the primitive forest; and everywhere an undergrowth of formidable nettles threatens the intruder. Great black ants torment the traveller with their dangerous bites, and crowds of innumerable insects follow and persecute him on his path. After having avoided or overcome all these obstacles, he arrives before massive ramparts of bamboo, thick as the arm and 50 feet high, whose hard, glassy bark turns the edge of the best hatchet. When this new obstruction is overcome, the traveller at last reaches the majestic dome of the virgin forest, properly so-called; gigantic trunks of bread-fruit-trees, and of the teak, the wood of which is almost as hard as iron; leguminous

plants, with their clusters of splendid flowers; bar-
ringtonias, fig-trees, and laurels, form the colonnades
which support the wonderful leafy vault.

Monkeys are sporting merrily from branch to
branch above him, provoking him by making him
the mark at which they throw their fruits; as he approaches, he sees the orang-outang with severe and
melancholy aspect, leaping from a moss-covered rock,
and with the aid of a club, making his way into the
thicket. The forests abound in animal life, unlike
our own silent forests of the West. Here climbing
plants rise spirally round the colossal pillars, and
overtop the gigantic trees, forming from the root to
a height of 100 feet, nothing but a single leafless
rope. The enormous leaves green and glossy, alternate with huge tendrils which support them, while
fragrant umbels composed of rich clusters of white
flowers, hang about in all directions. This plant, of
the family of the Apocyneal (*Strychnos tiente*), furnishes in its roots the terrible *rajah upas*, or poison of princes.

A tiger having received the very slightest wound
with a weapon that has been dipped in this poison,
or struck with a little wooden arrow blown through
the tube called the sarba-cane, begins to tremble,
stands on his feet for a minute, and then tumbles
over as if struck by lightning, and dies in convulsions. Curiously enough, the part of this tree that
rises above the earth, is harmless, and even the sap
has no dangerous properties. As the traveller advances, he meets with a splendid tree, the trunk of

which rises free of branches, from 60 to 80 feet high, bearing aloft a superb crown of foliage, which overtop the humbler vegetation around. But woe is the traveller, if his skin touches the milky juice which its bark contains in abundance, and which it is ever ready to spurt forth. Blisters and ulcers much more painful and terrible, though not unlike those produced by the poisonous sumac-tree, appear at once. This is the *autjar* of the Javanese; the *pohon upas* (poison tree) of the Malays; the *ypo* of the inhabitants of Celebes and the Philippines. (*Antiaris toxicaria.*) The use of this upas poison, a practice which at one time prevailed through nearly all the islands of the South, is giving way to European firearms everywhere except in the most remote and inaccessible parts of these islands.

THE VALLEY OF DEATH.

We cannot conclude this short sketch of poisonous trees, and especially the description of the upas-tree of Java, without saying a word about that valley the deadly character of which is attributed by the ignorant natives to the exhalations that rise from these terrible trees. Let us here again follow the narrative of Schleiden.

Leaving the dense virgin forest the traveller ascends a small hill and suddenly there is spread out before him a fearful wilderness, a genuine realm of death. The small level valley shows not a trace of vegetable life; nothing is seen but the bare soil burnt by the sun's

fierce heat. Death alone dwells here, and the ground
is literally strewn with skeletons. Often a tiger may
be seen lying as destruction seized him in the very
act of leaping upon his prey, or an unclean bird as he
swooped from the clouds to seize upon the carcass.
Piles of coleoptera ants and other insects are lying
here and there and bear witness to the justness of the
name, Valley of Death, which the natives have bestowed upon this place of desolation. But the fearful

The Valley of Death (Java).

character of the valley does not arise from the tree,
as was formerly believed, but from the emanations of
carbonic acid which, because of its specific gravity,
does not mix with the upper layers of the atmosphere.
As in the famous Grotto del Cane near Naples, and
the Vapor Cave at Pyrmont, in Germany, here also
this gas brings infallible death to all living beings
that breathe near the ground. Man alone, to whom
God has given the faculty of walking upright, trav-

erses with impunity this deadly valley, since these
choking vapors do not reach to the height of his head.
Just as the oppression felt in ascending the Himalaya
to the height of 15,000 or 16,000 feet, was attributed
by the natives to the poisonous emanations of certain
herbs, so the terrible phenomena of the Valley of Death
also were formerly charged to the exhalations of the
upas-tree, and assumed all the more formidable proportions in the mind of credulous natives and ignorant travellers, as no antidote has yet been discovered.

We do not envy the inhabitants of the tropics
their cow-tree, and, content with the gift of useful
caoutchouc, we readily give up all the rest of the luxuriant vegetation of those countries, which combine
such terrors with all their beauties. As yet no antidote known is able to counteract the effects of any of
these poisons, which as so many dismal enigmas
threaten the human race. They confirm the saying
that the brilliant light of tropical nature necessitates
equally dark shadows, and that more than one formidable dragon as yet guards the entrance of these gardens of the Hesperides.

The Oaks.

CHAPTER IX.

LONG-LIVED TREES AND GIANT TREES.

1. *Longevity of Trees.*

OF all the objects with which nature has clothed the earth, nothing gives us a clearer idea of the age of our globe than those trees whose branches have sheltered generation after generation for thousands of years. There is something mysterious and fascinating in the form of an immense, great tree. For our own part we have rarely witnessed the new life of spring, clothing, year after year, a tree with its splendid, ever-new apparel of leaves, without being deeply impressed and almost overwhelmed with the comparative brevity of human life. The monuments of man live longer than man himself, to be sure; but they do not live with the life of nature. The mountains also have witnessed the changes of ages, but they have no individuality with which we can become familiar.

The tree, on the contrary, like the flower, is an individual which watches us and stands before our eyes as a silent witness of our existence. That tree existed long before we ever saw the light—it has seen the ages that preceded us; men without number have passed at its feet who were our distant ancestors during these mysterious epochs previous to our existence. And when we shall pass away, and the place of our habitation shall know us no more, the tree will remain, calm and silent as to-day; it will put forth its leaves every spring, and other generations will come and play around its foot as we have done.

Large trees count their age by centuries. Who has not heard of the "Oak of the Partisans" in the department of the Vosges, which a few feet above the ground measures 40 feet in circumference and 16 feet at that part of its trunk where it sends out its main branches? Its height is 101 feet, its diameter 75 feet. It is nearly 650 years of age, and dates back to the days of Philip Augustus, when "partisans," as the rebels of the day were called, laid France waste.

At the base of the southern slope of Mont Blanc, in the forest of Ferri, near the pass of that name, there is a larch 18 feet in circumference above the collar of the root, which, by its size, has been presumed to be 800 years of age.

Not far from this larch is a pine on the mountains

of Béqué, which is called by the inhabitants of the country " the stable of the chamois," because it serves as a shelter for these animals during winter. It measures nearly 24 feet in circumference and 15 feet at the first branches. In spite of its magnificent growth and still verdant foliage, it is said to be 1,200 years old.

At first sight it seems astonishing that the appearance of a tree should enable us to determine, at least approximately, its age, and yet the explanation is very simple.

Every year a new layer of wood forms itself upon the tree, and when the trunk is sawn through the number of years it has lived is indicated by the number of concentric rings; since each ring in the wood represents the new layer which has been formed that year. A tree that shows a hundred rings, is generally regarded as having lived a century. It is by these observations upon the trees themselves or upon others of the same species, and by ingenious deductions, that botanists succeed in determining the age of trees.

The plants which in all parts of the world acquire the most remarkable size are the yews, the chestnuts, several bamboos, the mimosas, the cisalpinias, the fig-trees, the mahogany-trees, the cypresses, and the western plane-trees. We do not speak of the race of gigantic trees in California, which surpass all others, and the dimensions of which will be given in a future chapter.

THE AGE OF SOME TREES.

At Fortingall, in Scotland, there is a yew-tree more than 3,000 years old. In France, at Foullebec (Department of the Eure), a yew measured in 1822 appeared to be 1,100 or 1,200 years old.

Adanson measured at Cape Vert a baobab over ninety feet in circumference ; and by comparing it with younger trees of the same species he was led to believe that this giant was 5,000 years old; but doubts have since arisen whether the principle of measuring by annual rings can be applied to this family of trees. Golberg, measured another which was 112 feet in circumference, and consequently must have been still older. But the most remarkable is the colossal pine of California (*Sequoia*), which rises to the height of 300 feet, and is thirty feet in diameter. The concentric layers of one of these immense trunks, if correctly measured, prove that their age is 6,000 years, which would have made him contemporaneous with the earliest dynasties of Eygpt.

In Europe the lime-tree or linden, seems to be capable of living the longest and attaining the most gigantic proportions. The linden-tree of Neustadt, in the kingdom of Würtemberg, is a remarkable instance. Its magnificent crown measures 400 feet in circumference, and its branches are upheld by 106 stone columns. The tree was an old tree in the year 1229, when a great fire destroyed the old town, and the new town was, according to a document still extant, built close to " the big tree." In the year 1558, the

Duke of Würtemberg surrounded it with four porches, and caused his armorial bearings to be painted upon two of these columns. At the top the linden-tree of Neustadt divides into two great branches, one of which was broken by a tempest in 1773, while the other at the present day is still flourishing and 110 feet in length.

The linden-tree of Freiburg, though of smaller size, is of historical interest. It has grown up from a branch planted on the day of the battle of Morat, beside the corpse of a young Freiburger who had died on the spot from over-exertion in hastening to cheer his native town by the welcome news of the victory.

The linden-tree of Villars-en-Moing, near Freiburg, is still older, for it was already famous in 1476, when this great battle was fought. Its circumference does not measure less than 40 feet; its height is 75 feet; and its crown is still a vast mass of almost impervious foliage.

After the lime-tree, the oak grows to the greatest size in Europe.

England possesses very remarkable specimens both for age and size. The following are some of the measurements:

The famous oak of Clipson Park is 1,500 years old, since the park, which belongs to the Duke of Portland, existed before the Norman Conquest. The largest oak in England is the oak of Calthorpe, in Yorkshire. It is 78 feet in circumference at its base.

The Shire oak, so called because it stood on a spot

where the counties of Nottingham, Derby and York met, and its shadow thus covered a portion of each, extends its foliage over an era of 780 square yards. The most productive of oaks ever known, was one in the county of Monmouth. It was cut down in 1810; the bark alone was sold for 200 pounds sterling; and the wood for 670 pounds. (These figures are taken from the *British Review*.) In the manor of Tredegar, in the same county, a hall 42 feet long by 27 feet wide, was floored and wainscoted with the timber of a single oak-tree taken from the park.

The oak of Autrage, in the arrondissement of Bedfort (Upper Rhine), one of the largest trees in France, was felled a few years ago. It was 15 feet in diameter at the base and more than 42 feet in circumference. The trunk alone produced 4,500 feet of saleable timber. This oak is believed to have been in existence in Druidical times.

It is not necessary to travel far from Paris to see a number of very respectable specimens of vegetable antiquity. Without going as far even as the forest of Fountainebleau, and on the road thither, it you stop at the station of Montgeron or of Brunoy, and make an excursion into the beautiful forest of Sénart, you will come to the little village of Champrosay, near which there is a "cross roads," at which eight roads meet. In the centre of this opening stands the old oak of Antein. The trunk is 18 feet in circumference; and the space covered with the foliage is 90 square feet. Many of its branches have been cut down—they were no longer useful as formerly, to bear

THE OAK OF ALLOUVILLE.

the evidence of the owner's power of jurisdiction, who used to hang culprits there.

Among the ancient and marvellous trees which excite the interest of travellers in the highest degree, the immense oak at Allouville, near Yvetot, must be numbered among those to which memory most frequently returns. Much has been said and written about this tree; and though the simple villagers that dwell around know nothing of all the scientific discussions of which it has been the subject, they regard it with pride and with tender affections. Their ancestors have sat beneath in its shade; and their own children are now playing around it, as so many generations have done before them.

It stands in the centre of a graveyard, and often peasants from all the country around come to kneel under its heavy branches and there to pour out to God their sorrows and their grief. Just above the ground it measures thirty feet in circumference, and twenty-four feet at a man's height. In the interior of the hollow trunk a little chapel has been fitted up, and above, as it were in the second story, a rustic hermit lives, while still higher in the tree a small belfry, surmounted by a cross, has been built and crowns the marvellous edifice.

This oak cannot be less than 900 years old. The interior was fitted up as early as the seventeenth century and the chapel dedicated to the virgin. During the Revolution ignorant fanatics, who delighted in destruction, attempted on several occasions to burn down this venerable historic monument; but the inhabitants of Allouville and its neighborhood, who regarded the old oak with sacred fondness, turned out in arms and protected it against the Vandals.

Let us hope that many generations yet will enjoy its broad and pleasant shade.

The aspect of this tree excites a deeper interest perhaps than that of many structures left us by races existing no more. There seems to be something peculiarly eloquent in this great tree, that year after year renews its youth, though it has seen as many graves close and open again as the cold and **silent** stones of ancient temples; and we know of no history that has touched us more deeply than the humble and pious traditions of kings and warriors who have rested under its shade, of troubadours who have sung **its** praise, and of the tempests that have raged against it without ever impairing its beauty.

One day on a pleasure tour returning from Caudebec to Yvetot, we went out of our way to visit this famous oak. What struck us particularly about it was to find that little else was left of the tree but the bark. It is entirely hollow from the root to the top, and the interior is lined with wood, carefully plastered and wainscoted, like a monk's cell or an oratory, and yet the tree is still as green as those of the forest near by, and bears every year abundant crops of acorns.

The old chapel oak of Allouville is a monument of comparatively modest pretensions as to antiquity compared with the oak of Montravail, which is not less than 1,500 and perhaps 2,000 years old. This oak, which stands in the court-yard of the farm of Montra-

THE OAK OF MONTRAVAIL.

vail, near Saintes, is without doubt the patriarch of the
forest of the Saintonge and indeed of the whole of
France. It belongs to the species of *Quercus longaeva*,
and its admirable preservation promises to bear the
burden of ages to come. It is crowned each year
with green and abundant foliage, perhaps for the two
thousandth time. On a level with the ground, its diameter is nearly 30 feet and its circumference over 80
feet. The spread of its branches is 380 feet in circumference.

The decayed part of the interior forms a hall nine
to twelve feet in diameter and nine feet high. A circular bench has been cut out of the live wood for the ac-
commodation of visitors, and around the table in the
centre a dozen people can dine comfortably. It is dec-
orated with a living tapestry of ferns and mosses, and
light is admitted by a window on the left and another in the door.

Of this tree also but little remains save the bark.
This is the fate of almost all ancient plants which lose
their pith, their heart and their wood, and continues
to subsist only by means of their outer skeletons.
Such is the case especially with willows. We were
lately exploring the banks of the Marne, under the
magnificent viaducts of Chaumont, when one of these
willows arrested our attention. There was nothing
left of it but a mere shell; the tree was hollow from
top to bottom. It was still flourishing, and besides a
thousand parasites, animal and vegetable, lived in its
countless cracks and crevices.

Even beeches are known to have reached an almost

fabulous old age in some parts of England.   Such are
the famous Burnham beeches, near Slough, which
for size and picturesque beauty are unequalled.   Like
most pollarded trees, their girth is enormous, and their
moss-grown roots are thrown out in curious contortions,
grasping the ground as if setting all storms at defiance.
Tradition says that Harold's bowmen were encamped
in the wood a few years before the Norman Conquest,
and that the Danes pollarded the beeches.   An English poet says of them :

" Scathed by lightning's bolts, the wintry storms,
    A giant brotherhood, ye stand sublime ;
  Like some huge fortress each majestic form
    Still frowns defiance to the power of time ;
  Cloud after cloud the storms of war have roll'd
  Since ye your countless years of long descent have told."

Switzerland, so remarkable for the variety of its
natural treasures, adds to its beautiful scenery, picturesque landscapes and matchless prospects, special beauties, and among these latter some of the most celebrated trees known in Europe.

On the banks of the Lake of Geneva stands the
mansion of Meillerie, and the rocks on which it is
built are divided from the water only by the road to
the Simplon.   A little distance further on you come
to Neuve-Gelle, which has one of the most famous
chestnut-trees of the world.   Ever since the fifteenth
century this chestnut-tree has given shelter to a modest hermitage, and no doubt it was at that time already
a tree of respectable age.   At present its base measures

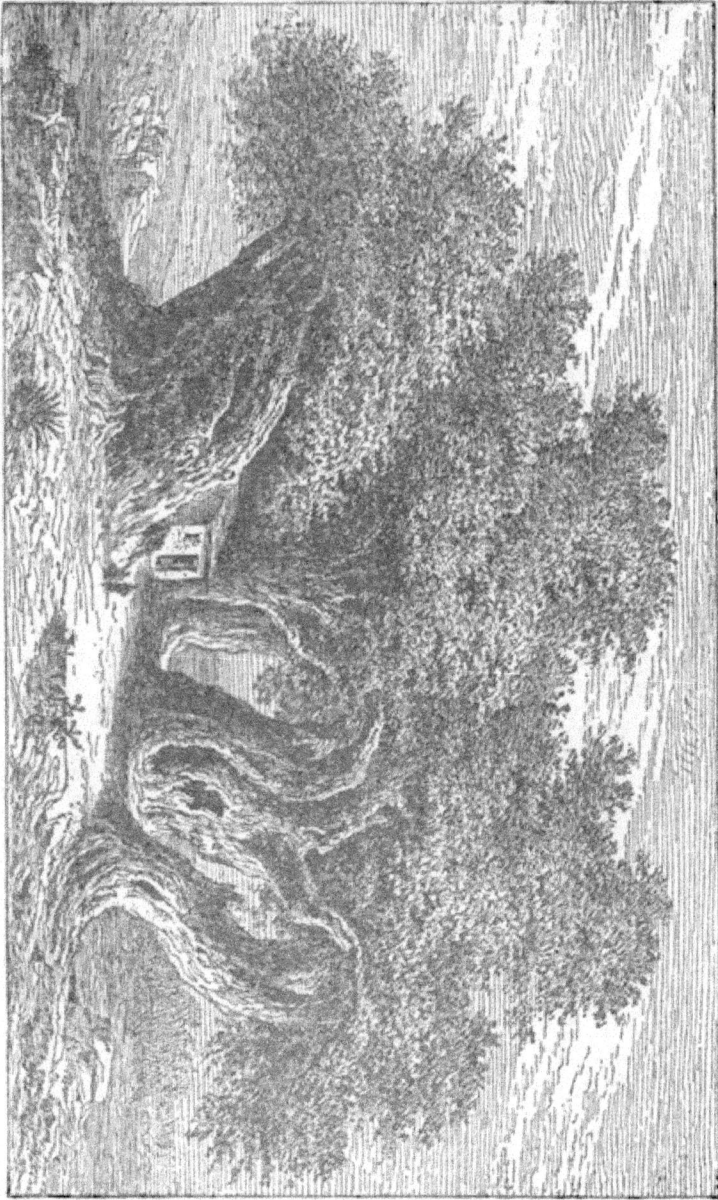

THE CHESTNUT OF MOUNT ETNA

forty-six feet in circumference; but having been re-
peatedly struck by lightning, it has been stunted in its
growth; the girth of its branches nevertheless gives
it a venerable aspect, and each summer crowds of
visitors come to see the famous tree and to rest under
its shade.

At Prilly, near Lausanne, there is a linden-tree
under which 500 years ago justice was administer-
ed. The municipality of Lausanne watches over its
preservation, dear as it is to the whole canton,
and a little fountain serves to keep its roots moist.
Nor must we forget the baths of Evian, where, a little
below the road grow two rose-trees of the same form
and almost equal in height and width. These are
not gigantic monuments, like the colossal trees of
which we have spoken, but they no less surprise travellers by their size. For these marvellous rose-trees are
certainly of respectable size for flowers—their trunks
measure more than ten inches in circumference.

THE CHESTNUT ON MOUNT ETNA.

The chestnut of Neuve-Gelle cannot compare with
this rival, under whose shade a hundred horses have
found shelter. It is said that the Queen Joan, of Ara-
gon, ascended Mount Etna during her voyage from
Spain to Naples, and that all the nobility of Catania accompanied her in her excursion. A tempest broke out,
but the queen and her whole suite found easily shelter
under the foliage of this immense tree.

" This famous tree of so vast a diameter," says
Jean Houel, the first traveller who gave a description

of it in the last century, " is entirely **hollow, and** subsists now only by means of its bark, **but does not** the less cover itself every spring with new foliage. The hollow of this tree **is** so large that the people of the neighborhood have constructed a house within, with an oven for drying chestnuts, filberts and almonds, and other **fruits** which they wish to preserve, **as is** the common usage in Sicily. When they require fuel, they take a hatchet and help themselves from the part of the tree around their **dwelling.** For this reason **this** magnificent chestnut **is nearly** destroyed.

" Some people believe that this colossus consists of several chestnut-trees, which, pressing the one against the other and no longer maintaining their individual bark, have grown together and appear as a single tree to careless eyes. This is a mistake. All the parts, though mutilated by time and the hand of man, belong to one and the same trunk."

Careful examination seems **really** to prove that all these diverging branches have but one system of roots. Moreover, Brydone, who visited it in 1770, states that his guide, following up the traditions of **the** country, assured him that at a time long past a single unbroken bark covered the trunk all around, although at the present day only a few remnants of it can be seen. Canon Recupero, a Sicilian naturalist, affirmed in the presence of the English traveller and many other witnesses, that the root of this colossal tree was a single one. The best proof in support of the oneness of this tree is the example furnished by **other**

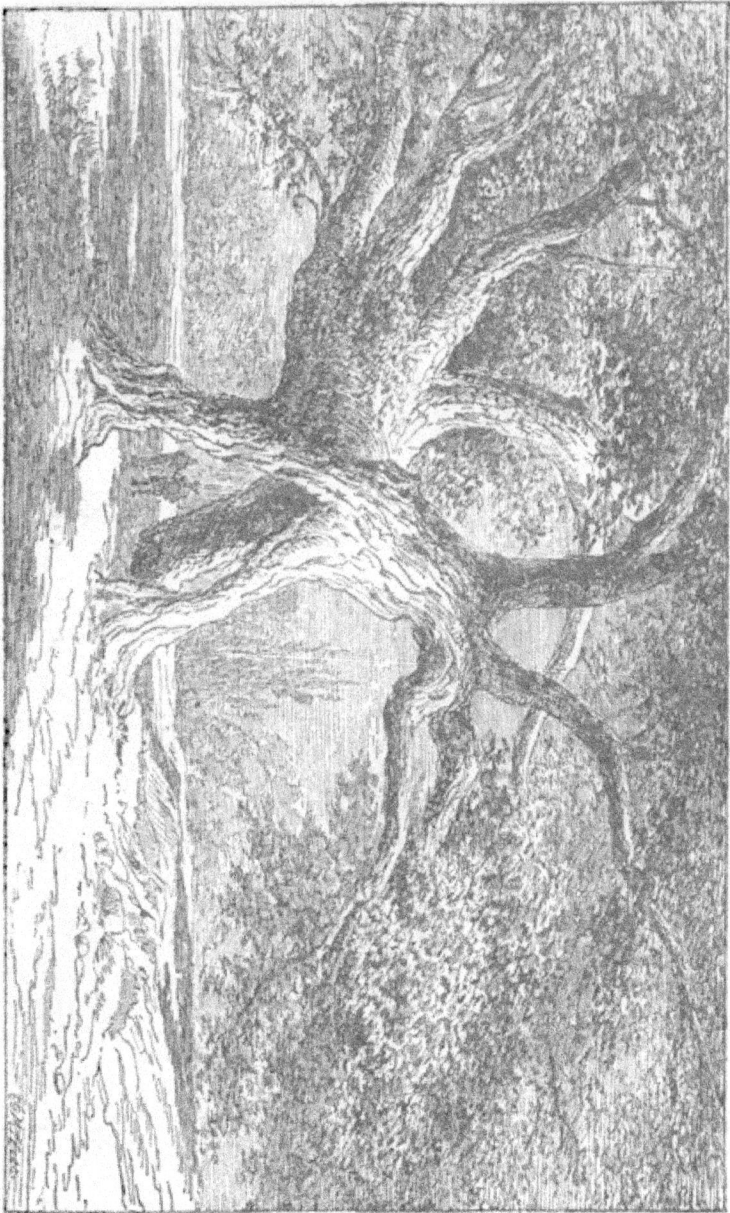

THE PLANE TREE OF SMYRNA.

chestnut-trees on Mount Etna, which have a diameter
of thirty-six feet.

The tree we are describing is 160 feet in circumference—it is impossible to assign a limit to its probable age.

At the present day an opening sufficiently large
to allow two carriages to pass through it abreast,
penetrates it from side to side, but this circumstance
does not prevent the venerable tree from covering
itself every year with bloom and fruit.

It ought to be added, however, in conclusion, that
it was the custom of ancient horticulturists to plant
around a single shoot a number of others of the same
species, so as to produce the appearance of a single
tree, which time would mature to a colossal size. They
peeled off the bark on the inside, and soon a single
bark came to envelope the whole. This practice was
pursued especially with olive-trees.

THE SMYRNA PLANE-TREE.

In the middle of the plain of Smyrna, in Asia
Minor, near the road that leads to Bournabat, is to be
seen the old plane-tree represented in our illustration.
Its singular form is not more surprising than its dimensions.

Bournabat is a village containing a grotto, in
which, according to tradition, Homer wrote the Iliad.
This picturesque place is the favorite retreat of the
rich merchants of Smyrna, who have built here their
country houses. But the pedestrians, and even the

horsemen travelling from the city to **the country, love**
to pursue **a** path running parallel **to and near the**
main road, which passes under this marvellous vegetable gate, formed by the divisions of the trunk.
The two stems, though divided, are sufficiently strong
to bear the enormous weight of the plane-tree, the
lofty branches of which afford a view **over** one of the
most lovely bays of the Asiatic coast.

From this point we can see **the** Oriental ceme-
teries of Smyrna, the most famous **next** to those of
Pera and Scutari, dark with the **sombre** shades of
countless cypresses. The view commands the plain,
also, from the eastern limits of the great city to the
fertile slopes in the west, that fall into the sea.

THE PLANE-TREE OF COS.

Cos, the celebrated **island of the** Sporades, that
gave birth **to** Hippocrates, **the** greatest of the physi-
cians, and **to** Apelles, the greatest of the painters of
Greece, contains in the centre of the public square, a
magnificent plane-tree, famous throughout the world.
Its far-spreading branches cover the whole square.
Left to themselves, these branches would break of
their own weight, if the inhabitants had not undertaken to support them on columns of marble. They
devote to this monarch of trees **a** kind of worship
not less sincere, nor less profound, than they pay to
the surrounding edifices, the last witnesses of their
former grandeur.

THE PLANE-TREE OF GODFREY OF BOUILLON.

What is called the plane-tree of Godfrey of Bouillon, is rather a system of nine trees forming three very close groups. Martins, who has seen and described it, regards it as the most colossal plant in existence; and Mr. Gautier calls it "not a tree, but a forest." "Commencing from the east," says the first of these writers, "we first perceive two trunks joined together, which, at a height of three feet above ground, measure 35 feet in circumference. A space of 15 feet has been hollowed out by fire. Then comes a single trunk 17 feet in circumference. The last group consists of six trunks, forming, so to speak, one tree, over 70 feet in circumference. This enormous trunk also has been hollowed by fire, for Turkish barbarism admires nothing, and respects nothing. A horse put up in this cavity, was perfectly comfortable in his new stable." Martins estimates the height of this vegetable mass to be over 180 feet. The space covered by the foliage is 340 feet in circumference. From the tents under the shelter of the tree, one can see Bujugdéré, a village on the Bosphorus, not far from which is the famous plateau of Godfrey of Bouillon.

THE YEW OF LA MOTTE-FEUILLY.

This yew-tree is at once the monument of nature and of history. A monument of nature, for it bears traces of an age that must be counted by centuries, its trunk being not less than 24 feet in circumfer-

ence, while the shadow cast by its still green branches, covers an extent of 70 feet.   A monument of history, for after seeing the Roman legions pass, it was
watered by the tears of Charlotte d'Albret, the unfortunate wife of Cæsar Borgia, Duke of Valentinois,
and by those of Jeanne, of France, divorced from
Louis XII., who came hither to mingle her grief
with that of her cousin.

At present one-half of the tree is dead, and no
longer reclothes itself with foliage in the spring ; but
the main trunk remains a permanent monument of
an age long gone by.   This yew stands in one of the
courts of the feudal castle of la Motte-Feuilly, not
far from the road leading from Châtre to Château-
meillant, on the borders of the former provinces of
Berry and la Marche.

THE ELM OF BRIGNOLES.

There is in the department of the Var, a little
river called the Caranci, which now flows outside of
the walls of Brignoles, but which formerly passed,
if we believe the local tradition, through the centre
of the square which still bears its name, at the foot
of a venerable elm.   This aged tree was already
well known in the fifteenth century, having witnessed
great events, and given shelter to countless guests.
In the sixteenth century, Michel de l'Hôpital sang
its magnificent proportions, to while away the time
of his exile in Provence.   King Charles IX. was
present on the 25th of October, 1564, at a *bal cham-
pêtre*, which was given under this gigantic elm

THE SYCAMORE OF TRONS.

Now this ancient patriarch of woods is supported by a wooden column, which is truly the staff of old age.

THE SYCAMORE OF TRONS.

In the long valley of the *Varder Rheinthal,* which shelters the Rhine in its infancy, stands the little town of Trons. Near the village, a venerable tree covers with its vast crown of leaves, a small chapel. In 1424, deputies of all the communities of the valley assembled under its branches in order to form the federation which resulted in the Republic of Grisons. The fourth centenary of this memorable event was celebrated in 1824, and in memory of the occasion, the little chapel was built, on the portico of which was the following inscription : " You are called to Liberty—Where the Spirit of God is, there is deliverance—Our fathers hoped in Thee, oh ! Lord, and Thou hast made them free." This tree was long called the plane-tree of Trons, and it is under this name that it is still known generally. It is, however, not a plane-tree, but a true sycamore.

At the elevation at which it grows, the plane-trees finds no longer the conditions under which it thrives.

At 20 inches above the soil, it is 28 feet in circumference.

In his journey to Nuremburg, Mr. Edouard Charton mentions a visit which he paid to the old linden-tree in that town, " planted," he says, " by the Empress Kunigunde." Formerly, on the occasion

12

of great festivals held in this German city, dances were held under its shelter. Its branches then covered the entire court-yard in which it stands. On the day on which Albrecht Durer's father took up his residence in the ancient town, in the year 1445, the patrician, Philip Pirkleimer, celebrated his wedding under this linden-tree. Four statues surround the tree now, representing four ancient emperors of Germany.

POPE'S TREE.

There is nothing remarkable from a botanical point of view, about the favorite tree of the poet Alexander Pope, near Binfield. It is a poor beach almost bare of leaves and branches, withered, and weak with age, and half destroyed by lightning.

Yet, in approaching it, a feeling of respect stirs our whole being. What a mysterious power there dwells in our association of ideas, which draw even inanimate things within the circle of our sympathies, and admit them, as it were, to the number of our friends.

Seven miles from Windsor stands the tree to which Pope came in his youth, to dream away the hours, and to receive his first impressions of the outer world. Its bark is covered with inscriptions in honor of the poet, and all around, on trees and stones, are engraved extracts from his principal works.

ROUSSEAU'S IVY.

In the English Park which surrounds the Italian Villa of Feuillancourt stands a gigantic poplar, which is entirely covered with a mass of ivy, growing larger and thicker every year. This originally sprung from a slip planted by Jean Jacques Rousseau, who was here as the guest of a gentleman to whom the estate at that time belonged.

The manner in which Rousseau put a sudden termination to the friendship between his host and himself is rather curious. The Duke of Noailles, proprietor of a very beautiful park at Saint Germain, wished to see and chat with Rousseau. As a direct invitation from the duke would certainly have been refused, for we know the misanthropic character of Rousseau, and his aversion to the world, the duke resolved to employ a stratagem, and requested the poet's host, Trochereau, to bring his friend into the park on the plea of making a botanical excursion. The duke was to wait behind the gate, and when the poet passed he was to appear as if by accident and invite both to come and see his collection of plants. All went well up to the moment, when the philosopher caught sight of the duke. In an instant he disappeared, and Trochereau sought him in vain. On the following day Rousseau wrote to his friend, informing him that from that day he would cease to know him.

There exists at Paris, in the Jardin des Plantes, a tree 230 years of age—Robin's acacia.

This plant, we are officially informed, has been the mother-plant of all the innumerable acacias which now adorn the gardens and woods of Europe and America. It stands in a square near the Rue de Buffon, but its worm-eaten stem, full of chinks that have been stopped with plaster, has been protected by an iron armor. As may well be imagined, nothing has been neglected that can prolong the existence of this ,patriarch of locusts, well known to all visitors of the ·Jardin des Plantes, who every spring eagerly come and examine its branches to see whether there is still life in the old plant.· But evidently its days are numbered. The sap, the life blood of the tree, circulates sluggishly through it, and a hundred signs proclaim that this tree, the oldest of all acacias, will soon be no more.

Placed at the extremity of the museum of mineralogy, in a part of the collection but little frequented, it does not attract the attention of visitors to the same extent as the famous cedar of Lebanon, though perhaps it is really more worthy of notice. It was planted in 1635—a century before the cedar—in the place where it still stands, by Vespasian Robin. The father of this naturalist had received it some time previously from North America, and hence its botanical name of *Robinia.* This was the same year in which the *Jardin Royal* was definitely established by an edict of Louis XIII., and of the trees that were in existence at that time, this is the only one that remains. It is also the first acacia that ever came to Europe. It has supplied not only France but Europe with one of the most useful as well as most beautiful species of trees. Not far from this acacia were formerly to be seen the first saphora brought from Japan, and one of the first horse-chestnuts from India that was ever seen in Europe.

The tree of the "Seven Brothers," in the forest of Villers-Cotterets is remarkable for its seven branches, which have been so disposed as to sustain a floor and gallery without injury to its foliage.

Walnut-trees live to a great age and reach occasionally gigantic proportions. One of the most marvellous is to be seen at Balaklava, in the Crimea. It produces each year 100,000 nuts, and is the property of five families.

The table of St. Nicholas, in Lorraine, mentioned by De Candolle, gives one an idea of the size to which these trees grow. This table is 24 feet wide and is all in one piece.

THE MAPLE OF MATIBO.

This plant, the type of the "Lower trees" with which the skilled horticulturist ornaments our gardens, is especially remarkable for its architectural form. This variety of the sycamore is not, in fact, a marvel of vegetation; and, strictly speaking, it is not on its own merits to be classed among the extraordinary plants which have already been described—plants which owe to nature alone this distinguishing feature. It is to be seen in its perfection at Matibo, in the vicinity of Savigliano, near Coni, in Piedmont. The skill and perseverance of the horticultural architect has made an astonishing metamorphosis of it. In its cultivated state it appears like a structure of two stories. Each of these has eight windows naturally formed and can contain twenty people. The flooring is managed by a skilful arrangement of the branches, which are carefully interwoven, and the leaves form a natural carpet. The birds of the air sing amid the the green leaves, and are not disturbed by the people that sit at the artificially formed windows.

More elegant than the oak of Allouville which we have already described, this sycamore does not appear to belong to the same category. It is only mentioned here as a type of the trees manipulated by art, with which gardeners decorate country houses.

THE TALLEST TREES.

In closing our description of the most remarkable specimens of the largest trees in nature, we have to mention in the first place the dragon-tree of Orotava.

"This colossal dragon-tree," says the author of the Tableaux de la Nature, "is found in the garden of M. Franqui, in the little town of Orotava, one of the most pleasant places in the world. When we climbed the Peak of Teneriffe, in 1799, we found that the circumference of this tree was about 45 feet some little distance above the ground. At the top it was nearly 80 feet, which, considering its girth at the surface of the soil, is not a little surprising. Tradition states that among the Gouanches this tree was an object of veneration, as the olive was among the Athenians, the plane among the Lydians, and the banana among the inhabitants of Ceylon."

In the year 1402 the dragon-tree of Orotava was accurately measured by the companions of Bethencourt, at the time when they discovered the island, and it was then as large and also as hollow as it is to-day

THE DRAGON TREE.

From this statement we might form a guess as to the age of this famous tree, remembering at the same time how slowly it grows. Berthelot says, in comparing the young dragon-trees in the neighborhood with this giant: " The calculation which we make as to the age of the latter inflame the imagination."

The dragon-tree has been cultivated from the remotest times, in the Canaries, Madeira, in Porto-Santa; and a very careful botanist, Leopold Von Buch, has seen it growing wild in the island of Teneriffe; it is not, therefore, as people have for a long time believed, a native of the East Indies. It is found at the Cape of Good Hope, on the Isle de Bourbon, in China and in New Zealand. Different varieties of this tree are found in these remote countries; but it does not exist at all in the New World. *Aiton's Dracœna* of the north is nothing more than a **Convallaria**. Borda measured the dragon-tree of the **Villa Franqui** in 1771. It is said that in the fifteenth century, soon after the Spanish conquest, mass was celebrated on a little altar that had been erected in the hollow of its trunk.*

The monumental character of these plants, and the degree of respect with which they are regarded, have made naturalists curious to ascertain their age and to measure more exactly their dimensions. De Candolle, Unger, and other distinguished botanists, do not hesitate to state that many dragon-trees now existing date back to the earliest periods of our history, to a time,

* Unfortunately, this famous tree was completely destroyed by a hurricane in the autumn of 1837.

in fact, when even the history of Greece and of Italy
was but just beginning. The sterility of these plants
is one cause of their longevity.

By the side of the dragon-trees, which in spite of
their enormous dimensions belong, strictly speaking,
to the same class as our asparagus, may be placed the
Adansonias or baobabs, which are certainly among the
largest and oldest inhabitants of our planet. The
earliest description of these trees is dated 1454, and
was written by a Venetian, Louis Cadamosto. He
found at the mouth of the Senegal a number of these
trees, the circumference of which was about 100 feet.
Perrotet says he saw baobabs over 30 feet in diameter.

Among the regions remarkable for their vegetable
products we must, in passing, mention the Island of
Tahiti—the Queen of Oceanica.

Without fully adopting the opinion of Bougain-
ville as to the magnificence of this island, and without painting the inhabitants in the glowing colors
used by Bernardin de Saint-Pierre, we must repeat
what has been so often stated, that for wonderful vegetation the South Seas are the most astonishing. The
natural productions seen in the islands of this region
make it the most famous in the world.

In Tahiti especially, the vegetable kingdom shows
the greatest perfection. " All along the coast," says
Prat, " grow in abundance the *Artocarpus incisa*,
Forester's Pine, the banana, the cocoa-nut, the *Spon-
dius Cytherea*, the *Pandanus adoratissimus*, the
paper-mulberry and others. In the interior of the

island grow mimosas, gigantic bamboos and palms.
On the slopes of the mountains grow in rare perfection those huge tree-ferns, so much sought after by
botanists. The greater part of our European vegetables have succeeded there—even the culture of the
vine has been attempted and grapes have been produced. Vanilla grows in perfection, and coffee and
the sugar-cane flourish so well that their export is
one of the chief commercial features, although in at-
tempting to raise these plants three serious difficulties
arise : the indolence of the natives, the excessive
price of labor, and a weed, called goyavier (guava-
tree), the roots of which are found everywhere. Nor
ought we to omit mentioning among these remarkable
trees peculiar to certain portions of our globe, the
mangroves, natives of Tropical America and India.
Long looked upon as strange hybrids, half tree, half
fish, living half plunged in the sea or the lagoons
near the coast, these trees (*Rhizophora gymnorrhisa*)
send down little rootlets from their branches, all ready
to start a new growth as soon as they sink into the
mud which surrounds the mother plant. Thus they
form a new family group around the parent stem, and
soon spread out into a vast, almost impenetrable forest, full of mysterious awe and exhaling a deadly
miasma.

The highest trees in the world are to be seen in
Van Dieman's Land. They are called in Australia
marsh-gum trees, divided into red, white and spotted
gums ; they belong to the genus *Eucalyptus*, represented in our country by the evergreen myrtle. The

dimensions of one of these trees are as follows : height, 300 feet (200 feet from the root to the first branch); diameter at the base, 28 feet. The wood is so hard that it cannot be worked with ordinary tools, and nails cannot be driven into it.

Another gum-tree was 95 feet in circumference, and three feet above the soil it would take 20 men to encircle it with outstretched arms.

The quantity of wood yielded by one of these trees is marvellous. The first we have mentioned weighed not less than 1,540,758 lbs.

These trees are the colossi of the vegetable world. They are to the oak and linden as the whale is to the elephant or the hippopotamus. They owe their peculiar name to the enormous size of their crowns, which are covered with dense green foliage. This variety of balsamic vegetable yields a highly esteemed oil in immense quantities, a gum which is sometimes eatable, and excellent wood for dyes as well as for cabinet work. Even among the *Eucalyptus* there is one species so large that it has obtained the special epithet of the gigantic (*Eucalyptus gigantea*). They passed for the tallest in the world until recent investigations in California resulted in the discovery of trees still taller. Even the baobabs of which we have spoken are exceeded in size by these California monsters.

Of late the Eucalypti have furnished various articles of commerce and become highly valuable. Their hard wood serves for furniture, their twigs for walking sticks, of which thousands are annually imported

PITCHER-PLANT

from Algeria. But the most touching service they render comes from their fragrant leaves. During the Franco-German war they were largely used in hospitals instead of lint, their balsamic nature not only curing wounds, but after a few hours causing all unpleasant odors to cease.

THE GIANT TREES OF CALIFORNIA.

California appears to be the country of the greatest vegetable wealth, as it is also the land where gold most abounds. It is about 15 miles from what was once French Gulch that we meet with the mammoth trees, which there count up to about ninety in number. In other places, and near the Yosemite Valley, hun-

Eagle Wing.

dreds have since been discovered. They rise straight as columns to the height of about 320 feet, and measure about a hundred feet in the circumference of their foliage. The first branches strike off at about

100 feet above the ground. They are not numerous, but the foliage is abundant at the top. One of the most beautiful and perfect of these trees was felled in 1855, and was carefully examined. From the results obtained, it appears that 4,000 years at least must have passed before it could have attained the proportions to which it had grown. Among the trees felled at this time one, a giant among giants, was found to be 450 feet high—higher than any cathedral or monument in the world—and 120 feet in circumference. In falling, this giant broke at 300 feet from the base, and even here it measured 18 feet in diameter.

The forest in which these giants grow is called Mammoth Grove. It is situated in a little valley near the source of one of the tributaries of the river Calaveras. The valley which produces these trees contains nearly 300 acres of ground, and lies about 4,000 feet above the level of the sea. During the summer the climate is delicious here, free from the choking heat of the lower plains. The vegetation is always fresh and green, while the water, as pure as crystal, is almost as cold as ice. The respective positions of these trees has led to their receiving different names. Of these the chief are the Man and the Woman, because these two stand together; Hercules, a felled tree which yielded 72,500 feet of timber; the Hermit, so named because of its isolated position; the Mother and the Son, etc. These trees have all a circumference of from 55 to 60 feet, and the height is in no case less than 300 feet.

GIANT TREES OF CALIFORNIA.

Several of these trees are 1,400 or 1,500 years old. One of them that was felled was so large that when its bark was transported. to San Francisco, where it was put together again to be carried to Europe, it was large enough to accommodate a numerous com-

Fallen Monarch.

pany, who put a piano inside, while twenty persons were dancing on the floor, and fifty occupied seats to witness the strange ball. The bark of one of these giant trees was transported to London in parts, the parts were set together again at the Crystal Palace, and kept there till a few years ago a disastrous fire destroyed the wing in which they had been placed. Our illustration represents the largest of this class of trees, which have been called the "Fathers of the Forest." "This tree, also known under the name of the mammoth tree, was found," says the botanist Muller, "by Lobb, upon the Sierra Nevada, at a height of 5,000 feet above the sea level, near the sources of the Stanislaus and St. Anthony rivers. It belongs to the family of the coniferæ, and reaches an average height of from 250 to 300 feet. Recent observations, however, inform us that it may attain a height of 400 feet and upwards. The diameter of its trunk at its base is from 12 to 31 feet. The bark, which is 18 inches thick, is of the color of cinnamon, and has on the inside a fibrous texture, while the wood of the stem is reddish in color, but is soft and bright." This reminds us that the wood of the baobab also is by no means very hard, although it is one of the most ancient trees in the world. About ninety of these immense pines are to be seen within a circuit of one mile. Usually they are grouped in twos and threes upon a black soil well watered. Even the anxious gold seekers have not been able to look upon them with indifference, and have called one of them the "Miner's Cabin." The stem of this tree is 300 feet high, and in it an excavation has been made 17 feet wide. The "Three Sisters" are a separate growth from one and the same root. The "Family" consists of two old and twenty-four young trees. The "Riding School" is a great tree hollowed out

THE FATHER OF THE FOREST.

by time, the cavity of which may be entered on horseback without inconvenience. It is astonishing that vegetable wonders like these should have re- mained unknown for so long a time; the fact might

Pioneer Cabin.

warrant us in supposing that even more gigantic trees may yet be discovered by men like Livingstone, and his fearless brethren, in all parts of the globe. We give in our illustration one of these gigantic

cedars (*Wellingtonia gigantea*), which was long considered the largest of large trees on earth.   -

It is a matter of extraordinary interest that these
giant pines of California, which were regarded with
awe by the aborigines, have caused one of the few
superstitious to crop up which are found among Amer-
icans.   The gold miners of that region, we are
told, sometimes tip a cone with the first gold they
find, and preserve and sometimes even wear it on
their person as an ornament, hoping that it will bring
them luck.   The curious belief has probably been im
ported by some of the German miners, in whose
homes superstitions connected with fir and pine are as
abundant as the crop of pine cones themselves.

THE MANDRAGORA OR MANDRAKE.

Country people can still recall the fright which the
simple name of this plant used to produce among
our ancestors.   It was a vegetable, but seemed also,

The Mandrake

by some mysterious association, to be something of a human being; and the works written on magic in the middle ages, numerous and implicitly believed as they were, ascribed to it a supernatural power. Pliny mentions it fully, though Theophrastes calls it first *Anthropomorphosis;* Columella speaks of it as the "half man (*semi-homo*); Eldal as the "man-headed-tree," and popular traditions all over the world, as the "little-man tree," etc. Its root is certainly a striking likeness of a headless man walking rapidly on footless stumps; indeed, much more like a man than any thing the ancient Mexicans have graven on their monuments, or our Indians write upon the leaves of their birch-bark mystery books. It was a principal ingredient in the composition of philtres, potions, and charms intended for evil purposes, and to cure barrenness. It is well known that the Septuaginta gave the name of mandrake to the plant dudaim, found by Reuben (Genesis, ch. xxx.), while Luther arbitrarily changed it into simple "lilies."

The man who found the precious root was deemed happy, since it gave him power to excite love wherever he chose, and exercised a divine influence upon his destinies; but its extraction was protected by weird, magic forms. When the little man-plant was uprooted, it gave forth groans; it had to be gathered under a gibbet, observing particular rites, and even when gathered according to rule, displayed its wonderful powers only under certain conditions. The best method, Josephurahead teaches, was to get a dog to uproot it, who paid with his life for his boldness;

and then carefully wrap it in a winding sheet.   Marvellous virtues were ascribed to it, not the least of which were, that it doubled the prices of money around which it was wrapped; while the Arabs still call it the " *Devil's apple*," from its power to excite voluptuous feeling.

This plant belongs to the family of the *Solonæ*, and its scientific name is *Atropa mandragora*. It is a poisonous plant, and grows in dark wood by river banks; and in those mysterious localities not often visited by the rays of the sun.   The root is thick and long, whitish in color, and generally forked like the legs of a man.   Oval leaves crown the root and spread all around, hanging downwards. Its white flowers are tinged with purple; its fruit, which resembles a small apple, has, like the whole plant, a fetid odor.   It is principally the bifurcation of the root which makes the plant look a little like a human body.   Another plant used for the same purpose, as mentioned by Schleiden, is a bulb (*Allium victorialis*, to which similar powers are ascribed.

In the same class with the mandragora, we must mention the ginseng of Tartary, discovered in 1616, by Father Lafitau, and presented by him to the Duke of Orleans—the Regent of France.   He relates his discovery in the following words:

" Having spent nearly three months in vain search for the *ginseng*, chance showed it me when I least expected to find it.   It was then in its maturity, and the red color of its fruit attracted my attention.   I at

THE DEVIL'S APP

once suspected that this must be the plant I was looking for. Having carefully uprooted it, I carried it
full of joy to a native woman whom I had employed
in searching for it elsewhere. She recognized it at
once as one of the simple remedies made use of in the
country, and explained to me on the spot its nature.
When I told her of the use made of this plant in China, this woman cured herself on the following day of
an intermittent fever that had plagued her for several
months. The preparation was simply a drink of the
cold water, in which a few roots of the plant had been
steeped after having been previously bruised between
two stones. She resorted twice again to this remedy
for the same complaint, and on each occasion she was
cured within twenty-four hours.

"My surprise was great when, upon hearing that
the Chinese name meant Likeness to Man, or, as the
translator of Father Kircher's work states it—men's
legs, I found that the Iroquois word *garentoquen* had
the same meaning! It signifies the two thighed, and
is by the Indians applied to the ginseng, the plant
which I had discovered in Canada and then again
known in China. Reflecting on the uncommonness
of the name, which seemed to rest entirely on the
very imperfect likeness to the human body borne
only occasionally by a few plants of this family while
it is met with in other plants of quite different species, I could not convince myself that the same word
should have been applied to the same thing in China
and in Canada without some interchange of ideas
and without direct communication. Thus I was con-

firmed in the opinion I had long entertained, that America had once been one and the same continent with Asia, and that it was then united to Tartary to the north of China."

When I had discovered the *ginseng* it occurred to me that it might be a variety of mandragora, and I had the pleasure of seeing that in this I was borne out by Father Martini, who says; " I cannot better describe this root than by saying that it is very like our mandragora, though a little smaller. It may be one of the species of that plant. For myself, I have not the slightest doubt that it has all the qualities and the same virtues as that plant. In shape the two plants are alike."

If Father Martini was right in calling the ginseng a variety of mandragora because of its shape, he was entirely wrong in calling it so on account of its properties. The European mandragora is narcotic, cooling and stupefying. These qualities are not at all found united in ginseng, yet Father Martini's ideas made me pursue my researches still further. I soon found out that the mandragora of to-day is not the plant so called by the ancients; and I believe that in examining the matter more closely and comparing ginseng with what the ancients called mandragora, we should find that the plant I discovered in Canada was the actual *anthropomorphos* of Pythagoras and the mandragora of Theophrastes.

It is easy to imagine how the mandragora of the ancients has been lost. In the first place it must have been in great demand in early times on account of its singular properties so highly esteemed by the ancients. Secondly, the difficulty that this plant has in multiplying itself would always make it rare. The roots would be constantly pulled out before the plants had come to maturity, and thus all chance of propagation would be destroyed.

The mandragora of the ancients being lost, nothing was more likely than that another plant should be substituted for it, which had similar properties. Our mandragoras have roots which bear some resemblance to the human body, from the waist downwards; the seeds are white, and have the appearance of small kidneys; and all these features are equally characteristic of the ginseng.

It need hardly be added that the reverend father's supposition has not been maintained by modern investigations. The ginseng of China, found abundantly in the Middle States of the Union, and largely exported to the flowery kingdom, is an araliacious plant belonging to the same class as our ivy. It is known to botanists by the name of *Panax Schinseng.* The plant found in Canada is an allied species (*Panax quinquefolium*), having a root like the ginseng of China. It is exported to China and highly valued there for its properties, which are however as fabulous as those ascribed to the European mandrake. There can be no doubt that the *atropa mandragora* is the plant of the ancient Greek writers.

Nympheacæ.

SECOND PART.

CHAPTER I.

FLOWERS:

AN attempt to describe marvellous flowers would involve a description of the entire flora of the globe; for, in truth, whatever its form, its size, its color may be, every flower is in itself a marvel of one kind or another. In order to follow some practical plan, we must first make a few general observations by way of explaining the plan we mean to follow, and afterwards we may choose some special types best suited to bring out into relief certain special features of the floral world.

The earth is a large garden sown with flowers, which add a singular charm to the domain that has

been given to man. In their succession throughout
the year they provide for us a continuous festival with
ever-varying decorations that follow each other in
regular order. First come cowslips with us, snow-
drops in Europe, long before the trees dare to put
forth their leaves. Then comes the crocus, timidly
peeping forth because it has but little strength to resist the strong winds; and with it comes the violet,
loved by all, and the bright primrose. These and a
number of wild flowers form the advanced guard of
the army of flowers, and their appearance, attractive
in itself, is all the more delightful because it announces
to us the approach of a vast multitude of beautiful
companions.

After these the children of nature appear in regular order and each month brings with it its own
proper decorations. The tulip begins to show its
leaves and flowers; soon the beautiful anemone will
spread out its purple-streaked petals; and the ranun-
culus display its magnificence, charming the eye with
its harmoniously disposed colors. The crown impe-
rial and narcissus, lilac and lilies of the valley, iris
and jonquilles, decorate the flower bed. At the same
time the fruit-trees mingle their soft colors with the
fresh, bright green of the early grass; and heighten
by contrast the beauty of our gardens.

At the same time the rose begins to show its
leaves and early buds, soon to claim the position of
queen in the world of flowers. No one can resist the
charms which it unfolds to view. The heart must be
stony that can remain without emotion at the sight of
14

a rose half opened in the rays of the morning sun;
glittering with dew-drops; and swaying gently upon
its slender branch as it is rocked to and fro by the
morning breeze!

Autumn in its turn brings its balsams and sun-
flowers, its tube-roses and *chrysanthemums*, its rich
carnations and brilliant colchicums, and a hundred
other varieties. Thus the glorious display continues
without interruption. Then comes sad winter with
its frosts, covering nature with a robe of snow, and
hiding it for a time from our sight; but while making
us long for the returning verdure of spring, it is busy
with those processes of regeneration under ground
without which there could be no floral display in
summer.

Let us pause here, and with Louis Cousin, reflect
upon the wisdom and goodness shown by this varied
succession of flowers.

How beautiful their combination of colors! How
diversified and how harmoniously are they blended!
What wonderful skill in the arrangement of these
tints! There, the colors seem to have been put on
with a delicate hand; here they are mingled according to the most learned rules of art. The color of the
background appears always to be chosen in such a way
as to bring out the drawings traced upon it, so that
the green, which surrounds the flower, or the shadow,
which the leaves throw down, serves still further to
give new life to the whole.

"In the flower," writes Pouchet, "this glorious and
supreme effort of vegetable life, the poetic imagination

of Linnæus beheld only the picture of a chaste marriage. Plants which are ornamented with visible flowers exhibit an endless variety of size, form, coloring, and perfume. While some, such as the valerians, bear such tiny corollas that we can scarcely make them out, the lilies and irises exhibit grand and sumptuous structures of this class, which rivet every person's attention; and yet some exotic plants leave them far behind in this respect.

The flower of one Aristolochia, which grows on the banks of the Magdalena, presents the appearance of a casque with great edges. The opening of it is so large that it will admit the head of a man; and Humboldt relates that, when travelling along by this river, he sometimes encountered savages wearing this flower on their heads like a hat.

But it is on the surface of rivers that the pomp of vegetation is displayed. Nature nowhere shows another flower which for size, united to coloring, can be compared to those of the Nympheæ and the Nelumbia. By gentle gradation they pass from the purest white to the most velvety red or the most delicate blue! In every age the magnificent plants have attracted man's attention and been the object of his admiration. Art has made a splendid use of them; and to them the ancient myths owe some of their most delicate and beautiful conceptions.

They play a great part in mythology and on Egyptian monuments. The colonnades of Thebes and Philoe, which seem to defy the hand of time, are crowned with capitals representing flowers of the

Nympheæ in full bloom, with which the sculptors of the Pharaohs have sometimes intermingled bunches of dates.

There is no Egyptian monument on which Iris is not represented surrounded by the lotus, or holding bouquets of it in her hand. This flower was the indispensable ornament of the immortal goddess. In the Hindoo temples it also serves as a seat for Bramah, who is represented sitting and holding in his hands the sacred Vedas.

Poetry has exhausted all its resources in telling of the perfume and color of flowers. Nature has surpassed art, and the pencil of Apelles and Reubens could not reproduce them in all their magnificence. And yet one color, black, is wanting amid this multitude of varied tints. Some corollas, such as those of certain Scabiosæ, are, it is true, of a sombre purple, but a perfect black is never seen in this organ.

One phenomenon occurs in respect to the coloring of flowers which has been a good deal talked about; it is the mutability of it. Pallas, when exploring the banks of the Volga, remarked with astonishment that a species of anemone, the *anemone patens*, sometimes bore white flowers, sometimes yellow, and sometimes red flowers. This phenomenon, still unexplained, appeared so abnormal that it was mentioned everywhere. It is, however, common enough; and we may observe it any time in France without encountering such a long journey.

The field pimpernel (*Anagallis arvensis*), so common in our country districts, frequently displays this change. Usually its flower is of a vermilion red, but it is also sometimes of a magnificent sky blue, which made some botanists think they were two different species.

A pretty little plant of the genus Myosotis, which is met with in our arid grounds, varies still more singularly in its color, for on the same stalk we find at the same time red, yellow, and blue flowers—a peculiarity to which this species owes the name of *Myosotis diversicolor* which has been given it.

Other plants display a still more remarkable phenomenon, for in them the same flower changes its colors at different hours of the day. This happens with the *Hibiscus mutabilis*, the corollas of which are white in the morning, become rose-colored towards the middle of the day, and in the evening take on a beautiful red tint.

The successive change in the tints of the corolla is easily conceived; it may depend on vital action or on chemical reactions affected by time; but what is much more difficult to explain is, that flowers having displayed a certain category of changes during the day, go through the same round of variation the day following. This is observed in the variously colored corn-flag (*Gladiolus versicolor Linn.*), the corolla of which, brown in the morning, becomes blue in the evening, and on the day following takes on again exactly the same succession of tints as it showed the day before. What a variety of perfumes the flower possesses! And yet notwithstanding their thousand and one shades of difference, those whose sense of smell is sharpened by practice can distinguish that of each species.

It is even stated in some works that a young American who had become absolutely blind, botanized, guided by the smell only, in the midst of prairies enamelled with luxuriant vegetation, and never committed any mistake in his gleanings.

Orchids.

CHAPTER II.

ORCHIDS OR AIR-PLANTS.

IT is not to the caprices of amateurs alone that orchids owe their celebrity ; they justify the predilection with which they are regarded by their beauty and their singularity, and even by the difficulties which explorers have had to overcome in order to bring them home from their intertropical forests, and by the care and skill which horticulturists have had to employ in acclimatizing them in northern climates.

In first speaking of their beauty and their singularity, we find that these remarkable plants have certain features utterly unlike those of all other plants. They live as parasites, either on the bark of large trees in equatorial forests—in which case they are called epiphytes—or upon the soil, in which case they are called terrestrial orchids. The first, by far the most numerous, hang from the shady vaults of tropical forests like graceful garlands of incomparable richness.

Here we are told they had adorned the brow of royal Indian maidens, flourished in the palaces and perfumed the luxurious air of Peruvian courts; they were the love-messengers of noble Mexican youths; they lay, a mournful tribute of affection, on the grave of a departed friend, and hung their fantastic garlands along the gold-glowing walls of Mexican temples.

"Under the tropics," says Humboldt, "the orchids enliven the trunks of the trees which have been blackened by the rays of the sun and the cliffs of repulsive rocks. The flowers of the orchids resemble sometimes winged insects and sometimes the birds which the perfume of their nectaries attract. The life of a painter would be insufficient to represent the magnificent orchids that grow even within a small space in the deep valleys of the Peruvian Andes."

Unlike any other parasites they enrich the tree on which they grow. Flowers of brilliant colors, infinitely diversified, decorate the upper branches of the trees and give forth a perfume so sweet as to become enervating. They grow downward, unlike other flowers, and seem to be purely aerial beings, the roots even finding their food in the atmosphere alone. The richness of their colors and perfume is such that not only Europeans admire and appreciate them, but the pow-

er of their beauty is fully felt even by the savage natives.

Another feature peculiar to these flowers, and not
less remarkable is, that like the country in which they
grow, they do not distinguish the difference between
the seasons and obey no regular law in their flowering.
They bloom apparently capriciously, and consequently
their flowers are seen and their perfume felt all the
year through. Besides, their bloom lasts two or three
times as long as that of ordinary flowers. The possessor of a collection of orchids is thus certain to have some
plants in bloom all the year round. Of course they
have to be kept in a green-house, the heat of which
is uniform throughout the year, and even besides this
they call for unremitting, intelligent, and minute care
in a greater degree than most plants.

The orchid which our illustration represents is an
acinctum, a plant recently introduced into France,
and very rare even in the most richly-furnished hot-
houses. The floral stem grows downwards, like that
of the stanhopæa and others ; the plant lives as a par-
asite upon trees and its flowers hang in low garlands
down the trunk.

These plants are still so rare in Europe, and with us,
that wealthy collectors have been known to pay fabulous prices for certain varieties. It is needless to
state that these enthusiastic purchasers are mostly Englishmen. Perhaps the most notable among them was
the Duke of Devonshire, who some years ago, when
visiting the collection of Mr. Henderson, was greatly
struck with the beauty of one of his orchids. The duke

was accompanied by a young lady passionately fond of
flowers, and both were charmed beyond expression with
the orchid.   But the collector could not be prevailed
upon to part with it ; it was unique in Europe and
was beyond price.   He was not proof, however, against
the munificence of the duke, who placed a pocketbook
well filled with bank notes in the hands of the astonished horticulturist and walked off in triumph with
his prize.

The terrestrial orchid is a native of Panama, and
a bulbous-looking plant.   From the base of the tree
a pale green, almost white, articulate flower spike
shoots up and rises to the height of four or five feet, the
upper portion forming a raceme of pure white, waxy
flowers, sometimes as many as twenty in number.
Each flower, waxy and pure white, is nearly circular,
about two inches in diameter, and in the centre the
column pollen masses with erect wings are so beautifully combined as to bear a remarkable resemblance
to a dove of purest wing, having the wings faintly
spotted with lilac.   In its native land this *Dove
Flower*, as the English call it, is known as El Espir-
itu Santo, and regarded with superstitious reverence as
a religious symbol at which no one who has ever seen
the flower will feel the least surprise.

Its only rival is the famous Butterfly Flower (*Onci-
drium papiteo*) from the verdant island of Trinidad.
This is an epiphyte or true air-plant, growing on the
trunks or branches of trees, to which it attaches itself
with great firmness by a network of fibrous and thread-
like roots, but entirely nourished by the atmosphere.

The flowerstalks, four feet long, support at the end a single flower, bearing a singular and striking resemblance to a butterfly on the wing, not only in general outline, but in some of its details, both of form and color. The centre of the flower seems a mimicry of the body of the insect; the sepals, long, narrow and slightly curved, represent in a wonderful manner in shape and position the antennæ of the butterfly, while the petals represent the wings, and the labellum or lip the expanded body of the insect. The striking and wonderful form of this flower, the brilliancy of its color, the position at the end of a long, neutral-tinted elasic, wire-like stem, when seen moving, we might almost say fluttering, like an insect, with every current of air, remote and apparently unconnected with any root or bulb, it requires no very vigorous exercise of the imagination to believe it to be not a flower, but a gayly-colored butterfly flitting among surrounding leaves and flowers.

But their number is almost endless; there is a spider-orchis and a bee-orchis, an orchis like a fly, one like a man, and another like a lizard. One is the very image of a swan, with arched ne.k and gently elevated wings (*Cynoches ventricosum*), while another (*Cabaestum viride*) opens a beautiful capote, with bows and strings complete, just the thing for a well-grown fairy on a summer's evening.

SCROPHULARINÆ

This elegant flower is the *Antirrhinum graccum*, and belongs to the family of the *Scrophularineae* (what villainous names for such pretty things). Few plants rival it in beauty and airiness. It comes originally from the Morea, and it seems almost to be a plant of the air, free from the weight and grossness of the things of this earth. It blooms in summer, and the flowers remain in full beauty for several weeks. The flowers, which are of a bright yellow, are very numerous and grow in bunches, while the leaves, with their graceful outlines, are alternated, and the stems are slender and beautifully interlaced with each other.

To this rich and varied order of plants belong a number of charming small flowers which adorn our gardens, and some of which are endowed with extraordinary medical properties. Such are the *Spudwell Veronica*, a bitter plant; the mullein, the hyssop, sharp and astringent, the fox glove, a poisonous plant of which only the very smallest doses can be taken with impunity, the cow wheat, the horse-wort, the pig-wort and the *paulownia*, all of them flowers and trees remarkable for their beauty and elegance. According to their species the flowers are either solitary and separate or united together in cymes, clusters or spikes.

Besides the *Antirrhinum graecum* which we have mentioned, there are other species not less worthy of interest—these are the wolf's mouth (*A. majus*), the snap-dragon (*A. aranteum*), and the *Antirrhinum augustifolium*, with its long and slender leaves.

The Yuccas.

CHAPTER III.

YUCCAS.

THESE beautiful plants, the palms of northern gardens, are now among the number of the choicest objects sought after by connoisseurs in horticulture. They are natives of this continent—well called the gorgeous wild lilies of America, and have only been seen in Europe within late years. Among their characteristic features are their leaves, which are in their way almost as useful as those of the papyrus, since they can be used for drawing and painting, as well as paper itself. They are very thick, very fine in texture and velvety, and serve for certain works of art, and for making light ornaments, fancy baskets, and artificial flowers.

The character of these plants recalls one of the most eloquent pages of the journal of poor Margaret Fuller, afterwards Marchioness D'Ossoli, in which she exhibited that rare union of deep sentiment with a conscientious study of nature. She is speaking of a man to whom society is no longer open, and who, like the illustrious prisoner of Fenestrella in the touching story of *Picciola*, had given himself up entirely to the study of nature, animals and plants.

"I had," said this person, "kept two specimens of *Yucca filamintosa* during six or seven years without their ever having come to flower I did not know the flowers of this plant, and had no idea of the sentiments which they might call forth.

In the month of June last I discovered a bud upon the plant that had the best exposure, and two weeks afterward, the second plant, which stood more in the shade, also began to bud. I imagined I was about to have an opportunity of examining the blooming of the one after the other; but no, the most favored of the two waited for his companion, and both of them bloomed together just at the time of the full moon. This coincidence struck me at first as uncommon, but when I saw the flowers by the clear light of the moon, I comprehended it. This plant is created for the moon, as the heliotrope is for the sun. It is not acted upon by any other influence, and does not unfold its beauties by any other light. The first night that I saw it in flower I felt a special delight, I might even say ecstasy. Many white flowers show to the greatest advantage in broad

THE YUCCAS.

daylight. The lily, for example, with its thick, firm petals of dead white, requires daylight to bring out its full beauty; but the transparent petals of the yucca, of a greenish white, and looking dull in broad daylight, gleam with a silvery brightness in the moonlight; nor is it only the plant which does not assume its true tint in the daytime, but the flower, which like all bell-shaped flowers, is unable to close entirely after it has once opened, contracts and nearly shuts up at noon, and lets its tiny bells droop sadly. The leaves, also, which at night seem vigorous and large, and stand out boldly from the stem in the shape of a fan, like the palm, appear languid and imperfectly formed during the day. Their edges seem ragged and unequal, as if nature, dissatisfied with her work, had left them without bestowing the last and finishing touches. On the day after the night on which my yuccas first bloomed I could not understand my misapprehension, for the flowers seemed to have lost all their beauty. But on the second evening I returned to the garden. There, in the soft light of the moon, my precious flowers expanded more lovely than before. The stem rose up into the air, straight as an arrow, all the flower bells grouped themselves around it in the most graceful way, and the petals, more transparent than crystal, shone with a softer light than diamonds; the outlines were clearly defined, and yet as airy as if they had been woven by the light of the moon. The leaves, which had appeared ragged during the day, seemed now bordered with the finest gossamer fringes. I gazed at my

15

beautiful plant until my emotion became so strong that I restrained it with difficulty. Then a thought filled my soul. It was that this flower of the moon was the most perfect symbol of beauty and of the purity of woman.

I have since had frequent opportunities of studying the yucca, and of ascertaining by frequent observations, the truth of what was revealed to me so poetically, viz.: that this flower blooms only at the time of the full moon, and that it veils its charms from the bright eye of day and reveals them only to the divine eye of the night."

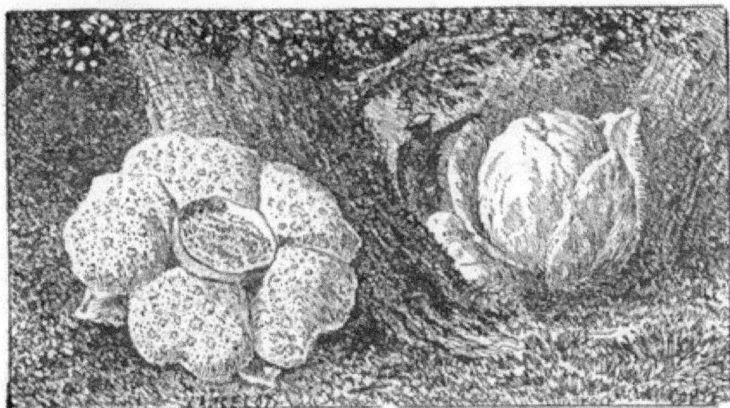

Rafflesia Arnoldi.

CHAPTER IV.

NYMPHEACÆ.—VICTORIA REGINA.—RAFFLESIA ARNOLDI.

D'ORBIGNY, the traveller, while travelling through the Republic of Bolivia, at Corrientes, was attracted by the flowers, leaves, and fruits, of a gigantic plant which he found floating on the water. This plant, one of the most beautiful in America, bears some resemblance to the water-lily, and seems to belong to the family of the *Nympheacæ*. The reader, in order to obtain some idea of what it is like, must imagine a vast extent of water covered entirely with huge round leaves, floating on the surface, measuring from three to six feet, with flowers now yellow, now violet, and sometimes white, more than a foot long, and diffusing a delicious perfume.

These flowers produce a spherical fruit, which, when ripe, is as large as an average-sized cocoa-nut, and full of round seeds, which are farinaceous. On account of this nutritive character the Spaniards have called the plant "water-maize," while patriotic Englishmen, impressed with the beauty and rarity of this colossal flower, have named it, in honor of their sovereign, Victoria Regina.

The *Lotus* is the sacred water-lily of the East, which appears in the mythology of almost every Oriental nation. In Egypt, where the flower reaches its greatest beauty, it appears constantly as the throne of Osiris, the god of day. In India, Vishnu was represented as a beautiful youth sleeping on a star-spotted serpent and holding the lotus in his hand. One of the holiest volumes of the Buddhists is entitled, "The White Lotus of the Good Law," and Buddha himself is always pictured bearing lotus flowers in each hand. The Syrians regarded it as a symbol of the cradle of Moses, found on the shores of the Nile by Pharaoh's daughter, and wherever the story of the deluge found its way the lotus was associated with the ark.

Our own water-lily (*Nymphæa*), growing in ponds and slow-flowing rivers, gives us at least a faint idea of the form and beauty of the Victoria Regina; but the South American plant is of gigantic proportions compared with ours. The large disks of round leaves, from five to six feet in diameter, are so many huge dishes of perfume. The leafstalk is below in the centre. The leaves are smooth and green above, with a rim

FLOATING LEAVES OF VICTORIA REGINA.

about two inches high, like the edge of a sieve or a large tray. On the under side they are reddish, and divided, into a great number of compartments by the veins, which project, leaving between them triangular or quadrangular spaces, each filled with a certain quantity of air by means of which the leaves are supported on the surface of the water. And so well do they serve this purpose, that birds and insects of all sizes may be seen walking upon these leaves or pursuing their prey across them as if they were on a solid surface.

Scomburgk, who discovered this flower in British Guiana independently of the traveller to whom we referred at the commencement of this chapter, lingers with pleasure over the description of this beautiful plant. The calyx consists of four leaves of a brownish red outside and white inside, each six or seven inches long and three inches wide. From these leaves of the calyx a considerable number of petals spread out in a circular and symmetrical form. These are white at first but become darker first at the centre and gradually turn to the color of the carnation. In many respects it is very like our water-lily. The petals, which are more than a hundred in number, gradually assume the form of stamens as they approach the central receptacle, which is fleshy, and bears large and farinaceous seeds on the surface.

Our indigenous water-lily presents, in respect to size, an appearance as worthy of attention as its exotic relative. It is in its way as richly-furnished with floral beauty as the Victoria. About seven o'clock in the morning it gradually rises from the water, and at midday it has attained a height of three inches above the surface. At four o'clock in the afternoon it begins to make its preparations for the night; it shuts itself up and returns gradually to its home in the water, where it remains till to-morrow's sun draws it to the surface once more.

In a memoir written by Ribaucourt we find some curious observations upon the developement of the leaves of this plant and on the prognostications which they suggest. It was without doubt by means of similar observations that Thalès, in the ancient times, gave a telling lesson to the inhabitants of Miletus. It had been objected to his science that it was unproductive, furnishing neither gold or silver. For an answer the philosopher bought up in advance the whole product of the olive-trees that grew around the town. He had predicted that the crop would be very abundant,—the result verified the prediction, and Thalès found himself the sole possessor of all the olives of the neighborhood. But content with proving thus that a philosopher can make a fortune as well as another, he distributed the whole of his gains among the merchants of Miletus.

The leaf of the water-lily sprouts from the end of its root early in autumn. It remains, however, very small, and rolled up during this season and the winter that follows; but in the spring it begins to grow and unfold itself as the season advances. Castel narrates that walking with a friend in September, 1788, along the shores of a lake abounding with water-lilies, he

was surprised to see that the leaves had already with-
drawn themselves under water, which they do not
generally do till about the end of October. From
this circumstance he augured that frost would soon
set in and that the winter would be long. The result
justified the prediction.

Certain vegetables are especially intended for the
ornamentation of different regions of the globe. The
Nympheacæ, floating on the surface of fresh, calm
waters, charm the eyes of wayfarers and painters in
all parts of the world. In Europe and in our own
country there are white and yellow water-lilies; Africa has varieties with blue flowers, and in the Indies
there are the *Euryale* and the *Nelumbium*.

Thus Egypt seems to have had its own peculiar
plant, the papyrus (*Cyperus Papyrus*), on the inner
bark of which they wrote. There are, it is true, various opinions entertained on the precise plant which
furnished them with their paper; but the majority of
naturalists incline to consider the paper-sedge the genuine papyrus of the Egyptians.

In like manner other plants appear to have a
special affection for certain mountain tracts. Such
are the rhododendron, charming shrubs with evergreen foliage, which adorn the middle regions of
shady slopes and which the traveller meets with either in the old world or in the northern portions of the
new world, blooming at the height of from 3,000 to
6,000 feet, now on the precipices of the Faulhorn and
now on the enchanting banks of the Lago Maggiore.

The beautiful flowers of this evergreen remind

us instantly of mountains and lofty Alpine regions, where snow-covered summits mingle with the white clouds. For the rhododendron marks the zones where the vegetation ceases and perpetual snow commences ; hence it does not flourish in the warm plain, and the *Rhododendron ponticum* cannot endure the rays of the sun.

The largest of all known flowers is one discovered in 1818, by Dr. Joseph Arnold, and described by Sir Stamford Raffles, then agent of the East India Company at Sumatra.

The first communication relative to this remarkable flower was addressed to the Linnæan Society of London, which immediately directed careful research to be made upon the subject and published the interesting results in its transactions.

This gigantic flower, surpassing all other parasites in size, was discovered during the first expedition of Sir Stamford into the interior of the province, an expedition in which he was accompanied by Dr. Arnold, member of the Linnæan Society, who would, no doubt, have realized the high hopes entertained of him had not death cut him off at the very beginning of his career.

Of this event Sir Stamford says in a letter :—" I am pained to have to report the death of Dr. Arnold. I had hoped, instead of sending you such sad news, to give you an account of the discoveries made by this youthful *savant*, and especially of the discovery of a gigantic flower, the largest, without doubt, that has ever been seen to the present day."

The following is an extract from a letter written by Dr. Arnold himself:

" Arrived at Pulo Lebbar, on the banks of the river Manna, I delighted in the anticipation of reporting to you that I had discovered what is, possibly, the most surprising prodigy that exists in the vegetable world. I had gone aside for some little distance, when one of my Malay servants ran up to me with astonishment marked on his features, and exclaimed: ' Come here, sir, come and see a new flower—very large, magnificent, extraordinary.' I went to the place to which the Malay directed me. "

And there to his astonishment he saw this colossus of the floral world. He had it cut and carried to his residence, where it was the astonishment of every one. It was examined, studied and copied, and it was from the sketch taken on that occasion that we give our illustration of the Rafflesia Arnoldi.

The five magnificent petals that spring from the centre are of a beautiful orange yellow ; in the centre of the calyx, upon a dark violet ground, rises a huge pistil, looking like a blaze in a bowl of punch. This prodigious flower is a yard wide, the petals are a foot in length, and stand at a distance of nearly a foot from each other. The nectarine or honey-cup looks large enough to contain a dozen pints, and the weight of the whole thing is given at 15 pounds.

Nepenthe.

CHAPTER V.

THE NELUMBIUM.—NEPENTHES.

AFTER the nympheacæ we will speak of the ne-lumbium, a class of magnificent herbaceous plants of a general conformation closely resembling the preceding, which grow in the fresh waters of the warm regions of Asia and North America. The flowers are very large and white, red or yellow in color. Besides the interest that is felt in the *Nelumbium speciosum*, the famous lotus of the ancients, two species call for attention, the brilliant nelumbo and the yellow nelumbo.

The flowers of the former are among the most beautiful and the largest in the vegetable kingdom.

They resemble magnolias, emit an odor of aniseed and grow upon long foot-stalks that raise them to the surface of the water. They are found principally in India and China, and are partly cultivated on account of the veneration in which they are held by the inhabitants of those countries. For the nelumbium is looked upon as a sacred plant—the symbol of fertility —and their divinities are represented as seated on its leaves. A species of the yellow nelumbio, grows in this country in South Carolina and Florida, but its flowers are smaller and always of a pale yellow.

THE NEPENTHES.

What Homer calls the *Nepenthe* has been interpreted as an allegory by Plutarch and some of the ancient writers already, because the flower now known by that name appears not to have been familiar to the ancients. The poet is believed to have described, under the form of a glorious flower, the charming Queen of Sparta, who made her guests forget how time passed, by her wondrous tale of love and adventure. It is certain that neither Lamark nor Brongni, nor Jussieu, have been able to class this plant among those now known. One believed it might belong to the orchids; another compared it to the rafflesias; a third calls it a "plant of uncertain character;" while still others make of it, in despair, a special family, that of the *Nepenthées*, represented in India by the *Nepenthes distillatoria;* in Madagascar by a special genus, characterized by the leafy tufts of its cups; in Cochin China, by the *Nepenthes phyllamphora;* and in

Java by the *Nepenthes gymnamphora.* The *Cepha-lotus* is the Australian pitcher-plant, and exceedingly curious and beautiful, being ornamented with stripes of red and purple, while the *Sarracenia*, also called the side-saddle plant, from a faint resemblance to a lady's saddle, is quite common in this country and abounds in the marshy regions of Virginia. All the nepenthes are evergreen creepers, and in the tropics climb up the trunks of trees to the height of thirty feet. This habit, and the long dark green shining leaves render the several kinds of nepenthes highly ornamental; but the curiously constructed and grace-fully formed pitcher which hangs by a long slender stalk to the end of each leaf, places them among the most singular and attractive objects in nature. Drawing their own water unaided from the wells of the dewy night, they fasten down the lid to keep it sweet and untainted by the wandering bedouins of the air, to be ready, at call, for the master's sole use.

It is commonly believed among the Indians of the mountains that if they cut off the little cups of a nepenthe, and empty out the contents, the day will not pass without clouds gathering and rain falling; hence, when they have reason to fear rain, they take good care not to touch this plant. On the other hand, when continued drought makes them anxious for rain, they hasten to turn over the cups or urns of the nepenthes. They hold this plant in high esteem, as being one of the most precious to the traveller, although it often happens that no nepenthes are found except on the banks of rivers, the waters of which are preferable to that of these vegetable cups in which tiny insects are apt to deposit their eggs.

"The structure of the cups appeared at first perfectly inexplicable to botanists," says a correspondent of the *Magasin Pittoresque ;* "for in no other plants are actual tendrils found to develope themselves in such singular manner." Dr. Hooker, the great English naturalist, however, soon ascertained by careful observation the manner of its developement. The leaf, as it first unfolds, presents a curious tendril or cirrhus, extending beyond the extreme point of the leaf. As this tendril lengthens, the small enlargement at the end increases, and the tendril, in the mean time, gradually bends upwards at the point like a hook ; the part thus bent continues to enlarge, the substance of the stalk appearing to swell, until it attains the size and form of a pitcher. The lid then separates from the rim excepting at the upper and outer side, where it remains more or less raised and united, as by a hinge, to the pitcher. This pitcher, being attached at its base to the slender, tendril-like stalk, hangs suspended six inches or a foot from the point of the leaf with which it is connected. Forty or more pitchers sometimes hang around a single plant. As the pitcher swells, and while the aperture remains hermetically closed by the lids, a quantity of pure, tasteless and colorless water collects in the cavity, which, when the lid is raised, is generally found at least one-third full of this infiltrated fluid. In the *Rafflesiana,* the somewhat pear-shaped pitcher is six inches deep and two or three inches in diameer at the base! The whole of the outside of pitcher

ter at the base! The whole of the outside of pitcher and lid is spotted with a rich, brownish-red or purple.

It is well known that similar modifications of the leaf stalk, and the leaf itself, occur in many other

Ice Plant.

plants also. Thus, in the water calthrop, which forces its roots down into the mud but spreads its graceful leaves over the surface of the water, the leaf-stalks are seen swelling in the middle into a kind of bladder, filled with air, which serves to sustain the plant. The leaf-stalks of the orange expand into real leaves, those of the mimosas often take the place of actual leaves, which remain abortive; and those of other plants, such as cherry-trees and apricot-trees, bear glands which correspond exactly to the cells that line the interior of the cups.

In the marvellously beautiful ice-plant of the Ori

Pyralis of Vine.

ent (*Mesembryanthemum crystallinum*), all the superficial cells of the leaves are so excessively developed
that they look like so many small bags filled with
limpid waters; hence the appearance of the plant,
which seems to be covered with drops of frozen waters and refreshes the eye in the midst of a dry and
dusty landscape.

In other plants the leaves seems to have a special
attraction for insects, which, leaving blossom and fruit
alike unharmed, are irresistibly attracted by a mysterious charm in the leaves. Legions descend upon the
unfortunate trees, among which the evergreen coniferæ alone harbor 400 species, all more or less hurt-
ful. One of these, the pyralis, destroys the leaves of
the vine and with them the life of the whole plant,
thus carrying despair to all the vine-growing regions
of Europe and our own country. For even in our own
vineyards the obnoxious insect has made its appearance, and, so far, science and experience have been
alike unable to contend with the feeble and apparently insignificant moth.

THE OUVIRANDRA FENESTRALIS.

In the conformation of its leaves this plant, a native of Madagascar, is not less remarkable than the preceding. Its leaves assume oddly enough the form
of windows, and hence the odd name the plant is
made to bear. The vascular network is left without
the diploe, which covers the leaves of all other plants
of this family. It is a vigorous plant, growing at a

depth of a foot or more under water; the root a large, oblong fleshy tubercle, out of which come forth cylindrical fibres. The leaves, which remain curiously enough always beneath the surface, are pistulate, elliptical, and pierced with innumerable holes in the form of parallelograms, and very close to each other. They consist exclusively of the elegant network of nerves, and thus present the appearance of delicate lace —hence the plant is also known as the lace-leaf plant. The color is bright green, and the whole leaf looks as if composed of fine tendrils wrought after a most regular pattern so as to resemble a piece of bright green lace or open needle-work. The flower-stalk, green and cylindrical, is the only part of the plant which rises above the surface and terminates at the top in two to five finger-like spikes, consisting of small pink flowers, exhaling a delicious fragrance. These and the seeds develop under the influence of light and air, which the leaves never seem to enjoy.

Among marvellous flowers we ought also to mention the Vallisneria, the typical species of which is the Vallisneria spiralis. The rivers of Southern Europe are adorned with numerous specimens of this remarkable plant, which was long looked upon as one of the most extraordinary in the whole kingdom of Flora on account of the marvellous phenomena which it presents at the time of flowering. The female flowers appear floating on the surface, as if in anxious expectation of others, which they are to fertilize. As if in obedience to their call, the male flowers, borne upon a long spiral stem, gradually rise from the bottom of the

16

pond, unrolling the long flower-stalk, turn after turn, till they also reach the surface. Here they meet the first-comers; they touch, and immediately begin to retreat once more to their dark homes beneath the waters, where they ripen their seed and provide for new generations.

We cannot very well leave this part of our subject without turning our attention to a phenomenon more general and more important than any to which we have yet referred—that of the migration of plants. Without this power of spreading abroad and actually moving from place to place we could not enjoy the richness of the natural carpet with which the earth is covered.

The learned director of the Museum at Rouen, M. Pouchet, shall be our guide here as well as in all questions of general import which require the assistance of a practical botanist. " Nothing," he says, " reveals to us more vividly the splendid resources of nature than the facility with which she covers the whole surface of the globe with vegetation and with life. This is attained not merely by the wonderful fecundity with which she endows plants,—she employs also the most ingenious and varied processes for transporting her fruits and seeds from one pole to another."

The vast number of seeds which certain plants bear ensures their continual reproduction; and upon this point calculation leads occasionally to unexpected results. Ray has counted 33,000 seeds upon one stalk of poppy and 36,000 upon a single stem of the tobacco plant. Dodard sets down at even a higher figure the number of seeds that can be gathered from an elm —according to him this tree furnishes every year more than 520,000!

It is evident that if all these seeds attained development, it would require only a few generations for these plants to cover the whole surface of the globe. But a multitude of causes retard this threatened invasion.

The fecundity of some mushrooms is still more extraordinary. Fries has counted more than 10,000,000 spores upon a single individual of the *Reticularia Maxima*. Other plants of the same family produce a still greater number of possible successors—indeed the abundance is so extraordinary that all the powers of the human intellect do not enable us to compute their actual number.

The majestic Arancaria of Patagonia bears at the tips of its branches 20 or 30 fruits of one tree, and each fruit contains about 300 kernels. Except by scattered families of the savage natives who subsist mainly on these fruits, the country is almost untrodden by man and left to itself, and hence the arancaria has formed, according to the interesting account of Dr. Peoppig, immense forests extending north and south for over 800 miles.

On the other hand, such is the fecundity of some of the gigantic *Lycoperdon gigantium*, that microscopic spores must be counted by millions of thousands of millions. But although they are invisible to the eye, each of these spores can produce a mushroom which, in one night, may attain the size of a large gourd! And it may be said without exaggeration that if the seeds of this plant were dispersed by a miracle over the globe, and were to spring up simultaneously, on the morrow the entire surface of the earth would be covered with hideous mushrooms!

The atmosphere is, of course, the chief agent in the dissemination of plants. A multitude of seeds have been furnished with feathery plumes, or with membraneous wings, for the sole purpose of enabling them the more easily to be carried away upon its current.

For this purpose the light fruit of many plants is surmounted by a plume of gossamer fibres, forming a real parachute which rises upon the lightest breath of the zephyr.

Borne away from the mother plant and mounting on the wind by means of this balloon, like a tuft of feathers, the seeds perform enormous voyages. The gentlest breeze bears them from the lowly valley to the highest mountain top; and if a tempest arises the frail parachute is whirled away by the storm, joins the clouds on high, crosses oceans and effects its descent in distant, unknown countries.

Other seeds, too heavy to be borne upon the winds, and able to endure soaking, accomplish long sea-voyages, and cross oceans by the aid of currents and waves. Thus cocoas, protected by their woody casing, and carried off by regular currents, pass from the shores of the Seychelles to the coast of Malabar, a distance of 1,200 miles. Astonished by this unexpected and mysterious phenomena, which is repeated every year, the Indians, as we have seen, can only explain it by supposing that the trees which produce these enormous fruits flourish in the unseen depths of the ocean.

It is, however, to the action of fresh water—to the currents of rivers and brooks that the most important migration of plants are to be traced. If Pascal has called rivers "roads that run," plants seem to have discovered the fact before him. Borne on their flowing waters, seeds frequently travel over great distances and find new homes in remote lands. Even at home land is continually washed away from river banks or shores and thrown up again elsewhere, full of tiny seeds.

Animals also contribute largely to the dissemination of plants. Bees and other insects do much planting; marmots, dormice and hamsters provision their underground dwellings with fruits, and a portion of their commissariat, often forgotten and left underground, germinate and develop at the return of spring.

Other mammiferous animals assist in their dissemination by a still simpler process; seeds mature in their fleeces and are deposited by them here and there in their peregrinations. Thus sheep are made to disseminate the seeds of agrimony.

If birds consume an enormous quantity of seeds, they are made useful, in return, by Providence, to assist, energetically in scattering other seeds broadcast over the land which they inhabit. Thus, for instance, thrushes, who feed upon the berries of the mistletoe, have been made to disseminate those celebrated plants throughout France. Other birds by the same means propagate in their turn a great number of plants. Travellers tell us, that when the Dutch had destroyed the nutmeg-trees in many of their East Indian islands in order to enhance the value of their nutmeg plantations in Ceylon, a variety of pigeons, which are particularly fond of these fruits, soon repeopled the localities with nutmeg-trees almost in every place where they had been apparently extirpated by the barbarism of the Netherlanders.

Even man is forced by Nature to do duty as an agent in disseminating plants. His vessels and caravans, traversing oceans and deserts, carry unconsciously seeds and plants and spread them abroad in new countries which he thus prepares, in blind obedience to higher powers, for his own future use.

NUTMEG TREE

Antirrhinum Graecum.

CHAPTER VI.

THE SENSITIVE PLANT.

"Weak, with nice sense, the chaste *mimosa* stands;
From each rude touch withdraws her timid hands;
Oft, as light clouds o'erpass the summer-glade,
Alarmed, she trembles at the passing shade,
And feels, alive through all her tender form,
The whispered murmurs of the gathering storm;
Shuts her sweet eyelids to approaching night
And hails with freshened charm the rising light.
Veiled, with gay decency and modest pride,
Slow to the Mosque she moves, an Eastern bride;
There her soft vows unceasing love record,
Queen of the bright seraglio of her lord.
So sinks or rises with the changeful hour
The liquid silver in its glassy tower.
So turns the needle to the pole it loves,
With fine vibrations quivering as it moves."

THUS the poet Darwin sings, in his fanciful
"Loves of the Plants," the praise of the Sensitive
plant (*Mimosa sensitiva* and *predica*).

Every one is, of course, familiar with the singular movement which the leaves of this plant show when they are touched. At the gentlest contact, they shrink back upon their supports; these fall back upon the common leaf-stalk, and the common leaf-stalk finally upon the main stem. If the extremity of one of the little leaflets be cut, the others close around it as if in sympathy. The leaves of this plant are digitate, that is, they are formed in rays branching off from a common centre, like the fingers of a hand. The narrow, straight leaves draw close to each other, as soon as they are touched, till their upper sides meet. They come together in the same way at nightfall, or when a frost is sufficiently sharp to affect the plant. In calm and warm weather, they are fully expanded; but when the plant is shaken by the wind, all the leaflets close simultaneously, and the leaf-stalks droop together. Even a simple cloud passing over the face of the sun, is sufficient to change their position, their expansion diminishing as light and heat decrease. Though closed, and apparently in a state of sleep during the night, they shrink still more closely together if any one touches them. At the junction of the petiole with the stem, and of each leaflet with the petiole, tiny glands are seen, which are the most irritable points. To touch these glands with the point of a pin, is enough to make the leaflet close. If the shock is sharp, all the leaflets make in succession the same movement, and close two by two in regular order. The whole leaf only begins to droop, when all the little leaflets have closed up, as if the main body could not fall asleep till all the members have been overcome.

If a little drop of water is delicately put upon the leaflet, De Candolle was not able to perceive any movement; but if a drop of sulphuric acid was substituted for water, the leaflets shrunk instantaneously, and drooped. The irritation is not merely local; as we have said, it communicates itself from one part to another. The power of contraction resides in the tiny round cushions placed at the points of junction, which form a kind of knee-joint spring, or hinge, and allows the stem to bend and lie down.

Certain experiments would seem to prove that these delicate plants can, to a certain extent, accustom themselves to a measure of excitement. Desfontaine observed this in carrying one of them in a cart. At the first movement of the cart it closed its leaflets, and all its leaves shrunk. But by degrees, as the cart rolled on, the plant seemed to accustom itself to its new condition; its leaves rose once more, and its leaflets unfolded. If the cart started again, after having stopped awhile, the delicate plant felt the influence, as at first; but after some time it seemed to recover once more from its fright, and showed again all its beauty to the day. It is now, however, considered more probable that the sensitive plant loses its strange sensitiveness from the continual irritation. The power of closing its leaflets is for a time destroyed by the repeated application of the external, mechanical agent, and restored again only after the plant has enjoyed some rest.

Other plants move when they are touched, but in a less degree than the sensitive plant. Such are the *Dionea*, the *Oxalis sensitiva*, and the *Onoclea sensibilis*.

From the time of Pliny we have been acquainted with the fact of the sensitiveness of certain plants to touch. This naturalist says, that in the environs of Memphis, there was a plant like the acacia, the leaves of which, arranged like plumes, shrunk when they were touched and rose again after a time. This is evidently a sensitive plant, though the precise species is not known. Pliny, however, only quotes Theophrastes.

PLANTS THAT MOVE SPONTANEOUSLY.

Desmodium Gyraus.

All created beings form after all but one great family; for it is the same spirit that ordained the creation of the whole universe; the same laws direct them; the same power sustains them; and all the children of our great mother Nature are brothers, bound to each other by indissoluble ties. From the mineral to the human being, the series rises by imperceptible degrees; the same features belong at the same time to all three kingdoms, minerals, plants and animals, forming in truth, the most perfect unity that can be conceived.

Among plants, those which in particular appear to possess qualities belonging to the higher, the animal kingdom, are the sensitive plants, which exhibit spontaneous motion, whether in the normal state of the plant or from accidental causes. The leaves of certain plants are endowed with a motion which may be termed revolving, because it follows a regular curve, and thus describes in the air a figure like a cone. The tendrils of the bryony, and of our garden cucumbers, are endowed with this perpetual motion, the duration of which depends to some extent on the temperature. These motions are not apparent, except under close and minute examination. This is not the case with the motions of the *Desmodium gyraus*. In this plant the leaves consist of three parts: a large terminal leaflet in the centre, and two smaller ones, lateral, and springing from the base of the former. Now, for the whole lifetime of the plant, by day and by night, in wet or dry weather, in the sun or in the shade, the lateral leaflets perform incessant little jerks, not unlike those of the second-hand of a watch; one of these rises a little distance, and at the same time the other sinks by as much; when the first sinks, the other rises, the motions being thus alternate and regular. They are the more rapid in proportion as heat and moisture increase, and are most evident when the sun's rays are striking upon the plant. In India, on the banks of the Ganges, where the plant is in full vigor, it has been observed that the leaflets perform sixty of these jerks per minute, and furnish us thus, as it were, with a genuine vegetable watch. The large leaf performs similar movements, but much more gently. This plant was discovered in Bengal, by Mrs. Mouson, a distinguished English botanist, who died during her scientific travels. The Indians were found to observe these motions with a sort of superstitious reverence, and attach to the plants supernatural powers.

We said above that the motions of these plants manifest themselves either in the normal state of the plant, or in consequence of occasional and accidental causes. The *Desmodium* is an example of the former; an example of the latter is furnished by the

DIONAÆ MUSCIPULA—*Venus's Fly-trap.*

This singular plant seems to have received from nature faculties far superior to those of other plants. It opens its pink lobes, the springs are set, and woe betide the insect that approaches incautiously. Instantly one of its leaves folds back upon the fly, which in vain tries to escape from the treacherous beauty; another has in the same way caught a small worm, it holds it fast and will not let it go. When we look upon these caprices of nature, we can hardly help being tempted to believe that she has given to these plants some powers analogous to those which we admire in animals. Like them, this plant has action, life, spontaneous motion. We find it possessed, in fact, of all that indicates purpose and will.

The first specimens of this plant were brought from South Carolina to Europe by John Bartram, in 1788, for the plant is native of North America. It is a pretty plant, bearing several elegant white flowers, while the leaves spread out close to the ground and terminate in two lobes joined to each other by a hinge, and surrounded at their edges with prickles. These lobes lie open, like the leaves of a book, and a liquid resembling honey is spread lightly over the edges, which attracts the unwary fly. Between the two lobes, just where they join, there are three sharp bristles, and as soon as a fly, or any other insect crawling over the surface, happens to touch one of the bristles, the irritability of the plant is excited, and the lobes, suddenly closing, imprison the insect—like a rat in a common gin. Its efforts to escape have only the effect of closing the curious trap more firmly. The prison doors do not open until all movement ceases, or in other words, till the insect is dead; then the lobes unfold and wait for another victim.

Another pitcher-plant, peculiar to the United States, is the Darlingtonia (*Californica*), growing on the Sierra Nevada, at an altitude of 5,000 feet above the sea. When fully grown it bears a most striking resemblance to the upraised head and body of an excited Cobra, with hood expanded, and preparing for a spring. The head is at right angles with the vertical, hollow body, and apparently presents no opening by which an insect could enter; under the place where the lower jaw would be, hang two large reddish appendages, like the "wattles" of a fowl. Flies and insects of every kind are irresistibly attracted by the large pitchers which this plant bears; they alight on the red "wattles," and then fly upwards into the tube; owing to a sudden twist in the neck of the pitcher, they are at once compelled to descend the hollow body, and never return alive. The old pitchers are generally full of dead flies, and as they soon split and rot, the ground around the plants is strewn with heaps of insects.

Thus we have here plants, bloodthirsty, cunning in the capture of their prey, and destroying animal life on a large scale. " What mysterious forces," exclaims the naturalist, Pouchet, "govern the life of a plant!" These beings, now so graceful, and now so imposing in form, adorned with dazzling colors, filling the air with the sweetest perfumes, are they left destitute of the faculties granted to the lowest of animals? There are two views on this subject, both equally guilty of exaggeration. One has been pleased to overrate what they call the inner essence of plants: the other is guilty of degrading it beyond measure.

The ancients were especially guilty of the first excess. Empédoclus did not hesitate to attribute to plants the highest faculties; and some of the successors of the philosophers of Agrigentum have not stopped here. The mysterious mandrake was considered by them a being possessed of the most exquisite sensibility. At the least wound, the little-man plant was supposed to give forth piteous groans. And those who dared go in search of it preferred employing ample precautions to withstand the dread it inspired, and to escape from its malignant influence and harm.

Nor are the crude notions of credulous antiquity unknown to our own day; on the contrary, they have often assumed a still more fantastic shape. Adanson, bold philosopher as he was, distributed souls largely among the plants; one he thought was not sufficient for each and so to each he gave several. Hedwig, a profound botanist, Bonnet, who was more superficial, and above all, Edward Smith, attributed to plants an exquisite sensibility, and even the possession of most delicate sensations.

These ideas have found in our day ardent defenders in two of the most celebrated *savants* of studious Germany, Von Martius, and Theodore Fechner. These men look upon plants as sentient beings, endowed with individual souls, and the latter has carried his enthusiasm so far as to form a kind of vegetable psychology.

The genius of Descartes had succeeded in making the masses believe that animals were nothing better than simple automatons, wound up to perform a certain numbers of actions. Going still farther, other naturalists, like the great Huler, the founder of vegetable physiology, were disposed to look upon plants as beings, subject to no other law but that of material forces. But neither extreme finds nowadays favor with men of science; they do not look upon the children of nature as mere machines, but they are as far from believing that they possess souls. The phenomena of plant life are still more or less an enigma; they cannot be ascribed to natural and chemical causes only, and yet they can as little be traced back to the power of a supreme and individual intellect. Only one thing is certain: they are subject to a vital force which controls all the springs of their existence · where this vital force disappears, life is at an end, and destruction inevitable.

All the savants, however, who have examined the
17

subject thoroughly, agree on this point, that plants
enjoy a life as active as that of most animals, and that
they show signs of more or less sensibility.   Bichat,
in his magnificent work on Life and Death, admits
this without hesitation.   Numerous experiments prove,
beyond doubt, that there is evidently in plants a degree of sensibility analogous to that of animals.   Elec-
tricity affects them, and narcotics paralyze or kill
them.   If sensitive plants are watered with opium,
they are put to sleep like men.   Prussic acid poisons
plants with as much rapidity as animals.   Let us
throw aside the antiquated ideas respecting vegetable
life; let us simply examine the phenomena, and we
must arrive at conclusions which are astonishing.

"Although the existence of nerves in plants may
be still doubtful," continues the same author, " yet it
is certain that the irritability manifested by sensitive
plants seems absolutely under the control of organs
which are analogous to those of animals, since they
are impressed in the same manner and by the same
agents as those of animals.

Among plants endowed with marvellous qualities,
one may be cited that has furnished powerful tools to
quacks and charlatans.   It is the *Anastatica*, or Resurrection Plant, commonly called the *Rose of Jericho*.
It is a truly marvellous sight to watch this plant, when
apparently dead and dry, assuming once more the
color of life as soon as the root is plunged into water.   Its buds swell with new life, the leaves of its
calyx open, the petals unfold, the flower-stalk grows
and the full-blown flowers are before us like the work

of magic. The *Rose of Jericho* is not a rose, but has
been placed by Linnæus in the first order *Siliculosa*.
Its earliest mention, perhaps, is in Jesus Sirach, ch.
xxiv., and ever since it has been connected in popular
superstition with the Holy Land and the life of our
Saviour. It grows in the sandy regions of Arabia,
Egypt and Syria. The stem divides at the base and
bears spikes of pretty white flowers, which change
into round fruits; when the latter are ripe, the leaves
fall, the branches grow hard and dry, and fold inward
so as to form a kind of ball. In autumn it is uprooted
by the storms and carried towards the sea; there it
is gathered and exported to Europe, where it is highly
prized on account of its hygrometric qualities. All
that is necessary is to place the end of its root into
water, and soon the plant is seen to begin a new life,
to develop its parts, and to unfold new roses before
the eye of the enraptured observer. When the water is removed, the spectator sees the magical plant
grow weak, close up its petals, and the leaves pass
through the last agonies of vegetable life and die.
In certain countries it is still believed that this marvellous rose blooms every year on the day and at the
hour of the birth of our Saviour; while pious pilgrims
to this day report finding it at every spot where
Mary and Joseph rested on their flight into Egypt.

The natives of Mexico attached the same marvellous qualities to their *Resurrection Plant*, which is
also found in California, on the Pacific coast. It has
a more remarkable recuperative power than any other
variety, and after drifting about for months, brown

and shrivelled, it requires only a few moments in a cup of water to expand to its original form and recover its color.   Still another plant, a Euphorbia, called Medusa-Head, blooms out in warm water, after being apparently dead.

The Bindweed.

CHAPTER VII.

THE SLEEP OF PLANTS.

WHEN the shades of evening spread over gardens and fields the plant sfold their timorous leaves, as if feeling some premonition of the darkness and cold that are approaching. We have seen how sensitive plants close up their leaflets as soon as the absence of their beloved sun has made itself felt, or when they are touched by a foreign substance. This habit, however, is not peculiar to these delicate plants only; it is a feature of a great many other plants, which invert their leaves at night in a manner entirely different from their normal arrangement during the day. Their appearance is so completely changed that it is often difficult to recognize them in their strange disguise.

This condition is what Linnæus, who discovered it in Sweden, terms the sleep of the plants, although this expression, borrowed from animal life, and applied to plants, does not mean the same thing as with animals—a state of repose and flaccidity; for during the night, plants are as stiff and firm as they are during the day. Linnæus, in order to verify the difference in the condition of leaves during the day and the night, used to deprive himself of sleep, for several nights, and descend into his garden to examine his plants. He soon discovered that it was the absence of light only, and not the intensity of cold, to which this phenomena was chiefly due; and this fact was of use to him in establishing upon better authority than heretofore the connection that subsists between light and the organization of plants. He next carried some of them into green-houses, where they were protected from all injurious influences, and ascertained that even thus sheltered, the plants yielded as submissively to the negative influence of darkness as their companions in the open air. He also found out that the difference between night and day is much more keenly felt by young than by old plants; and constant observation proved to him that the object of nature in establishing this difference, was to provide for the early closing of the young and tender leaves, which are more sensitive than those of older growth to the influence of cold and the night air.

The positions assumed by leaves during the night depends much upon whether these leaves are simple or compound ; in the latter, the difference is far more distinctly marked. In the oxalis, with compound leaves, the leaflets bend toward the common stalk, and lean against it with their under surface, leaving only their upper surface visible. Sweet peas and common beans fold their leaves up, till one supports the other; while other plants roll theirs together, in the shape of an ear trumpet.

The common chickweed (*Stellaria medica*) furnishes a beautiful instance of the sleep of plants. Every night the leaves approach each other in pairs, so as to include between their upper surfaces the tender rudiments of young shoots. But they are not alone. If one were to walk in a botanic garden after the setting of the sun, a great number of plants would be found which present a different aspect during the night from that which they present during the day. In some the leaves are erect and cover the stem, in others they hang down and cover the leaflets with the under side ; while, in still others, they approach each other in such a manner as to form tiny boats. In the mallows, with simple round leaves, the form of the latter is convex or concave, according to the hour of the day.

To what cause are these general phenomena due ? They seem to be independent of the thermometric or hydrometric condition of the air. De Candolle, following the example of Linnæus, ascertained that light was the most direct cause. He exposed plants that closed their leaves at night to an artificial light, little inferior in brilliancy to that of day. "When I exposed these plants to light by night, and placed them in obscurity by day," he says, "they opened and closed their leaves at first without any fixed rule; but after a few days they adapted themselves to the new condition of things, and accepted night for day and day for night; opening their leaves with regularity at night, which now brought them light, and closing them during the daytime. When I exposed them to continuous light, day and night, they had, as in the ordinary state of things, alternate seasons of sleeping and waking; but these seasons were somewhat shorter than in nature. When I exposed them to continual darkness, they also slept and remained awake alternately, but the intervals were very irregular."

The natural inference from these facts is that this tendency towards periodic motion is inherent in the plants; and that light, acting with different degrees of intensity upon different species, is the chief cause of it. It must be added, however, that other botanists have failed to obtain the same results as Linnæus and De Candolle; so that the question is not yet absolutely decided. It is claimed by many naturalists, that there exists a hidden bond which connects the life of plants with the great luminary in the heavens.

CHAPTER VIII.

THE FLORAL CLOCK.

THE flowers of the lovely nipplewort, the beautiful water-lily, and the brilliant marigold, with a great number of other plants, expand and close at certain fixed hours. They mark the altitude of the sun, and its inclination; and steadily following its motion on high, by their own imitative changes on earth, they indicate, with unerring accuracy, the course of time. Having observed this remarkable fact, Linnæus contrived his famous floral clock. It consisted of three divisions: a meteorological division, containing flowers that open or close earlier or later, according to the condition of the atmosphere, and consequently indicate the state of the weather; a

tropical division, as he called it, consisting of plants
that open at sunrise, and close at sunset; and a horo-
logical division, consisting of flowers that open and
close at fixed and invariable hours. It is this last division that forms specially the floral clock. The following twenty-four flowers open successively at different hours of the day and night.

Midnight.	Large-flowered Cactus	Cactus grandiflorus.
1 o'clock.	Lapland Sow Thistle	Sonchus Lapponicus.
2 o'clock.	Yellow Goats beard	Tragopogon luteum.
3 o'clock.	Great Pieris	Pieris magna.
4 o'clock.	Smooth Hawks beard	Crepis teclorum.
5 o'clock.	Day Lily	Hemerocallis fulva.
6 o'clock.	Shrubby Hawkweed	Hieracium fruticosum.
7 o'clock.	Sow Thistle	Sonchus oleraceus.
8 o'clock.	Mouse Ear	Hieracium pilosella.
	Pimpernel, or Poor Man's Weath-er glass	Arragallis arvensis.
9 o'clock.	Field Marigold	Calendula-arvensis.
10 o'clock.	Neapolitan Mesembryanthemum	Mesembryanthemum Nea-politanum.
11 o'clock.	Lady Eleven o'clock	Ornithogatum umbellatum.
Midday.	Ice Plant	Mesembryanthemum crystal-linum.
1 o'clock.	Proliferous Pink	Dianthus prolifer.
2 o'clock.	Hawk-weed	Hieracium.
3 o'clock.	Dandelion	Leontodon Taraxacum.
4 o'clock.	Alyssum	Alyssum alystrides.
5 o'clock.	Evening Primrose	Œnothera biennis.
6 o'clock.	Geranium	Geranium triste.
7 o'clock.	Naked-stemmed Poppy	Papaver nudicaulis.
8 o'clock.	Erect Convolvulus	Convolvulus rectrus.
9 o'clock.	Linnæan Convolvulus	Convolvulus Linnæi.
10 o'clock.	Purple Ipomea	Ipomea purpurea.
11 o'clock.	Night-flowering Catch-fly	Silene noctiflora.

Among the flowers that open at a fixed hour, several do not open again after closing, as the Syrian
mallow; others, like most composite flowers, open
again on the following day.

A great number of flowers open only at night. Among the most remarkable of these is the large flowered cactus (*Cactus grandiflorus*) or, night-blooming cereus, originally from Jamaica and Vera Cruz. Its magnificent flower expands and diffuses a delicious perfume soon after sunset; but it remains open only a few hours, and before dawn breaks it has closed. Generally it expands once more on the following evening, and this continues during several days. For four years in succession a plant of this species opened its flowers in a garden in the Faubourg St. Antoine, on the 15th July at seven o'clock in the evening, with unfailing regularity.

Among the flowers which open and diffuse their perfume only at night we may mention the Arabian jessamine, several species of the *Cestrum œnothera*, the lychnis, several libnes, some geraniums, and a variety of gladiolus. The *Belles de Nuit*, our marvel of Peru (*Mirabilis Jalapa*), owe their French name to this habit of not opening till evening in hot weather.

The African marigold opens constantly at seven o'clock and remains open until four o'clock, if the weather be fair. If it does not open, or if it close before that hour, it is certain that rain will fall during the day. In like manner the Siberian thistle remains open all night, unless it is going to rain the following day.

The flowers of the nymphæa or water-lily, close and sink into the water precisely at sunset; they rise again to the surface and expand as soon as the sun reappears. Pliny mentions this fact: "It is reported," he says, "that in the Euphrates the flower of the lotus plunges into the water at night, remaining there till midnight, and to such a depth that it cannot be reached with the hand. After midnight it begins gradually to rise, and as the sun rises above the horizon, the flower also rises above the water, expands, and raises itself some distance above the element in which it grows." According to some writers, this circumstance is the origin of the worship by the Egyptians of the nymphæa lotus, which they considered sacred to the sun. Its flowers and fruit are often to be seen engraven on Egyptian and Indian monuments. The flower ornaments the head of Osiris; Horus, or the sun, is likewise represented seated on the flower of the lotus. Hancarville has proved it, that they considered this flower an emblem of the world as it rose from the waters of the deep.

In speaking of the floral clock, it may not be out of place to give the calendar in which each month is represented by its favorite flower.

January,	Black Hellebore.
February,	Daphne.
March,	Alpine Soldanella.
April,	Wild Tulip.
May,	Dropwort.
June,	Red Poppy,
July,	Centaury.
August,	Scabiosa.
September,	Alpine Cyclamen.
October,	Chinese Hypericum.
November,	Ximenisia.
December,	Cluster Lopezia.

PAPYRUS.

The Flora of the Sea.

CHAPTER IX.

MARINE PLANT.

SALT water covers nearly two-thirds of the surface of the globe. Is this immense extent destitute of the wealth and beauty of life, while the earth is endowed with such a wealth of animals and plants? The ancient naturalists were far from comprehending all the abundance of life in the ocean. Linnæus, even, speaking of marine plants, only mentions an insignificant number.

Science, more advanced in our day, has sounded the depths of the ocean, and in those dark regions has found an exuberance of vegetable life not inferior to that which the dry land presents. There is a world of its own, beneath the waves, and the classifications of land plants does not apply to those of the watery world. The sea-bottom is laid out in mountains and valleys, covered with a magnificent vegetation, in the midst of which a thousand animal forms are sporting—forests that shelter guests more numerous and not less varied than those of our more familiar forests on *terra firma*.

It is our duty, however, to state that if there are incomparably more animals in the water than on the earth, vegetable life is not so extensively represented in the former; but there is this compensation, that in the ocean there is still another class of creatures, which are at once animals and vegetables.

Yes, the sea is a new world, the rich and varied productions of which will hereafter form a most marvellous section of Natural History. The posthumous work of Moquin Tandon (The World of the Sea. London: Cassel, Petter & Galpin), has revealed the importance of this hidden world, and contains, as in one casket, all the pearls concealed beneath the waves. Let us hear what the great German botanist, Schleiden, says about submarine plants: " The submarine flora consists almost exclusively of algæ or sea-wrack. These plants present such a diversity of forms that a promenade at the bottom of the sea would not be less interesting or less varied than a journey in the Tropics. Their peculiar structure, soft and gelatinous in all its parts, a collection of organs round, elongated or flat, which do not deserve the name of trunks and leaves; their brilliant colors, green, olive, yellow, rose and purple, and sometimes combined in the most extraordinary way in the same plant, give to these vegetables a strange and fairy-like character."

"The plants of the ocean," says a French writer, "do not much resemble those which adorn our woods and valleys. In the first place, they have no roots. Those that float are globular or egg-shaped, tubular or membraneous, but show no signs of root; those that are stationary are fixed by a sort of gummy, superficial matter, more or less lobed and divided. The earth counts for nothing in their development, for their origin is always independent of it. Every thing takes place in the water—from it comes every thing— to it every thing returns."—(*Quatrefages.*)

Terrestrial plants choose particular localities and flourish only in certain soils. Marine plants are indifferent as to what rock they attach themselves, whether it be calcareous or granite, to them it is all the same. They grow indiscriminately anywhere— even on corals or shells. They have neither real stems nor real leaves; they spread out in wide or narrow layers, in one or many pieces, which supply the place of these organs. They sometimes resemble waving ropes, and at other times crisp threads. Some of them might be taken for little transparent balloons, for cakes of trembling jelly, for tanned hides, or for fans of green paper. Their surface is sometimes soft, polished, luminous, at other times covered with papillæ, warts or real hairs. Their color is dark or olive, yellow dark brown, dark or bright green, pink, or more

18

or less vivid carmine.   Some writers divide them
according to their color into three great sections; the
brown or black, the green and the red.   The first,
found always at some depth, are by far the most numerous, and constitute the greatest part of submarine forests.

The green plants are superficial, and often floating.
The red plants are generally found in shallow places
and attached to rocks near the shore.

Islands of weeds of immense extent, floating on
the surface and sometimes carried by currents to prodigious distances, are often met with by voyagers.
Columbus encountered one on his first voyage to
America.   These are formed of sea-wrack.   But at
the bottom of the ocean are rich fields of tufted plants
and of shrubs, where the fish, the bird of the sea,
builds his water nest, groves and gardens where
the inhabitants of the ocean sport, woods and forests
which afford hiding-places for the timid, unarmed
denizens of the sea to escape from the assaults of the
monsters of the deep.

One fact worthy of remark is, that submarine,
like terrestrial plants, attach themselves to certain
geographical limits.   When we consider that these
conditions of vegetable distribution are heat and
moisture, and remember that at a relatively inconsid-
erable depth the sea in all parts has the same degree
of heat, we will be astonished to find so many varieties
in the submarine flora in regions that are contiguous.
It may be said, however, that the algæ display the
greatest exuberance in temperate zones, diminishing

in this respect as we approach either the poles or the equator.

But at the bottom of the sea vegetation is richest under the equator. "Let us leave," says Schleiden, "the aquatic forest of the north and their gigantic plants, some of which, as the pear-bearing algæ (*macrocystis pyrifera*) are from 500 to 1,500 feet long, and turn to the regions where the sun is more powerful, to see if we find here the same profusion of vegetation. Let us plunge into the limpid crystal of the Indian Ocean, and immediately before our eyes will be displayed the most enchanting, the most marvellous spectacle. Massive trees with singular branches bear living flowers. Large and compact meandrines and astreæ form a strange contrast with their jointed arms covered with finger-like branches. The colors surpass description. The freshest green alternates with brown or yellow; deep purple tints blend with bright red, pale brown and the deepest blue. Some millipores, of a bright red, yellow or peach color, cover the withered masses and are themselves covered with beautiful pearl-colored retipores, resembling the most exquisite carvings in ivory. By their side delicate fans wave to and fro, the light yellow gorgoniæ, and the pure sand of the bottom is marked with stars and extraordinary forms of the most varied colors. Around the flowers of the coral, little fishes, reflecting a metallic sheen in red and blue, the humming-birds of the sea, sport like the spirits of the abyss, and medusæ steer their huge, milky white, or light blue bells across the enchanted region. Isabelles (*Holocan-*

thes ciliaris) and coquettes (*Lepidopus argyreus*) and a thousand silvery fishes, displaying the most glorious colors, abound everywhere and mingle with each other in the most wonderful manner, until a slight breeze springs up, then the mirror is broken and the enchanting scene disappears as if by magic.

At night the astonishing scene opens once more, but with the addition of strange phosphorescent illuminations and with still more dazzling colors. Millions of star-like medusæ and microscopic shell-fish dance up and down in the faint darkness like fire-flies. Further on the magnificent Sea Pen (*Veretillum cynomorium*) waves about in a magic light, fairer than her brilliant red in daylight, and everywhere sparks flit across the waters, fires blaze up and softer lights are diffused. What in the light of the sun looked brown and plain, now assumes all the tints of the rainbow, and, as if to fill up the measure of all that is grand and glorious in the dark deep, a gigantic moon fish (*Orthagoriscus mola*) passes by like a huge disk of molten silver, surrounded by thousands of sparkling stars. We will add but one feature. The solitary traveller who has examined the wonderful coasts of Ceylon, returned one evening richly laden with treasures to his dwelling. " All at once in the middle of the quiet night, lighted by the silver brightness of the moon, a sweet music, like the wild harmony of Æolian harps, struck the ear. These melancholy sounds, sufficiently loud to drown the breakers, came from the shore close by and recalled the songs of the Syrens. The music was caused by the singing muscles, which chant a sweet and plaintive melody from the coast."—*Athenæum*, 1848, *No.* 1089.

If we complete this panorama by a picture of the watery world of plants, where there is neither leaf nor calyx, nor corollas, and of the animals dwelling there, rich in colors like flowers and shining like stars ; if we consider the ever-changing mutability of the bottom of the sea, which by turn overflows and again abandons the continents of the world, we shall be able to form some idea of the power, the importance and the wealth of this element, which the eloquent poetry of the East has apostrophized as the first and eternal source of all things.

Forest of the Carboniferous Period.

CHAPTER X.

PLANTS OF PRIMITIVE TIMES.

THE vegetable carpet which in our times embellishes the surface of the terrestrial globe and yields fruits and flowers, has not always existed under the form in which it presents itself to us now. There was a time when the aspect of vegetation was essentially different, and the happy man to whom it might be given to survey the two vegetable systems of primitive and present times would be called upon to admire two worlds very different in their conditions of existence. In the primitive time of which we speak, no tree, bush, or flower, at present existing, was to be found on the face of the earth. The world presented a spectacle in every way different from ours.

There were, it is true, dense forests with shady

foliage, silent retreats and grand avenues, as at present; the wind sang among the branches; the rays of
the sun fell upon the morning and the evening mists,
and the whole of nature was full of life and movement.   But there was no human being to contemplate these glories, to listen to these harmonies.   It
is doubtful if the first representatives of animal life
had yet awakened into life in the depth of the ocean
or by the marshy banks of the rivers.   Plants held
universal dominion; the earth was a " vegetable kingdom "—and nothing more.

It would be a mistake to fancy that this primitive
vegetation consisted of plants larger, stronger and
more beautiful than those that clothed the earth
when the reign of man began, and it would be an
equal mistake to imagine that those ancient plants
were as rich and luxuriant as those we see around us.
At the time of the coal formation of which we speak,
probably not a single fruit or flower had yet appeared upon the earth; and as to the supposed colossal
size of these plants, let us see in what this comparative superiority consisted.

The beautiful trees we have described, the giants
of California, the monstrous baobabs, the elegant
palms, the gigantic oaks, and the brilliant and odor-
ous flowers of our own day had not yet emerged from
the mysterious birth-place of beings.   The earth
hitherto had only seen plants of great simplicity and
poverty of form.   The plants have at the present day
but a few rare representatives, which are not apt to be
noticed by the side of richer modern forms.   Every

one knows our marsh-plants, horse-tails and other
reeds, consisting of a single stem, cylindrical, hollow,
worthless and uninteresting; our lycopodiums or club-
mosses, our ferns, and all the host of modest, unsightly
cryptogams—these are the descendants in modern
times of the plants of the era of the first coal formation.   For the number of vegetable productions,
however, this period, the period of transition from the
primitive to the secondary epoch, is far superior to all
others; and to this amazing fertility we owe the un-
measured extent of the valuable coal fields which are
to supply our race for ages in Europe and America.
Instead of rising to the height of only one foot,
these " horse-tails," etc., rose then to the height of
40 to 50 feet, and the degenerate club-moss (*Lycopo-
dium*), which now reaches rarely three feet, grew in
primitive ages to a height of 90 feet.   In those
ancient forests club-mosses had the proportions of
stately trees.   Mushrooms sometimes attained 40 feet
in diameter, and tree-ferns, such as shown in our illustration, rose uniformly to a height of 30 feet at least.
But imagination would go greatly astray if it fancied
that in like manner our oaks measured then 200 feet,
our pines 400 feet, our elms 60 feet in diameter, etc.
" The young earth," says Zimmerman, " expended all
its strength in developing reeds and brakes, mosses and
mushrooms, and while we find mosses equal to trees,
and perhaps mushrooms as big as mountains, there
did not exist a single tree in those days larger than
those of our own times."

But although at that period the whole surface of

the earth was covered with vegetation, these few species had a very monotonous appearance. The modern
naturalist who could behold the earth as it then was,
would be struck by the vast expanse of forests which
covered the earth wherever the water had receded,
but also by the melancholy uniformity of the trees
forming these interminable woods and the absence of
all life. Not only were there only a few of the 200,000
species in existence which we now admire in their
matchless variety of form and color, but the diversity
produced by climate, from the tropical heat near the
equator to the eternal ice of the Polar Sea, was wanting, since climate itself was as yet an unknown feature
of our globe. The heat of the sun had little effect
by the side of the intense heat of the earth itself.
Hence, even now the fossils of animal as well as vegetable life of those days are invariably the same,
whether they are found in the Arctic Zone or in the
Tropics. One vast uniform forest literally covered
the whole globe. The heat at the poles, drawing its
power from the internal heat of our earth, was then
at least equal to the highest temperature now known
in the Torrid Zone.

Besides the simple horse-tails and ferns, of which
the humble representatives surviving in our day
give us a better idea than any design could do,
the primitive world possessed a few equally simple
varieties of plants, which have since entirely disappeared from the surface of the earth. Zimmerman
assures us that there are no plants now in existence
like those extinct vegetables.

But all these plants have been found in a petrified state in the rocks of the coal formation. There they are preserved for us in the most wonderful museum in the world. It is astonishing sometimes to find that the texture—the fibres and the pulp—have all preserved their forms unaltered, though the substance itself has entirely disappeared. The Town Hall of Nordhausen, in Germany, contains a staircase of sandstone, each fragment of which clearly indicates that it has been originally of wood. But no example is so remarkable as the *forest of petrified trees* which Sir James Ross visited in Van Dieman's Land, although it must be borne in mind that this forest belongs not to the first coal measures, but to the series of Tertiary strata.

"One of the most marvellous natural curiosities," says this traveller, "which attract geologists to Van Dieman's Land, is the valley of petrified trees, a great number of which are transformed into the most beautiful opal. While the exterior presents a bright, homogeneous surface, like a pine stripped of its bark, the interior consists of concentric layers, which appear perfectly compact and of the same nature, but which can be easily split up in the direction of their length. These trees are standing upright, and it would seem that they were in full growth when the burning lava overwhelmed them. Some fragments of this wood have been carefully examined, and looked so full of life, so absolutely like wood, that only a very careful examination brought the conviction that they were really stone."

Coal was formed, as we know, by the prodigious exuberance of primitive vegetation that covered the whole earth. Every one has observed that in damp cellars, in which dry wood is kept during winter, there is a soft wood layer left behind, which resembles vegetable mould; and it is also well known how our marsh-plants are gradually converted into peat. In a similar but infinitely more powerful manner was our early vegetation converted into coal.

At the time that the vegetable world was preparing for man the fuel necessary for his industry, it appears to have been called on to play an important part in the economy of nature—that of purifying for the good of the aërial creatures who afterwards came to exist) the atmosphere which was surcharged with carbonic acid gas. For though this gas is of great importance to the growth of vegetables, it is an obstacle to the existence of animals, and especially to the more perfect classes of animals, such as mammals and birds. But when that ancient and abundant vegetation fell and was closed over by the earth, the carbon was no longer mingled with the air, which gradually became purer and better suited to the existence of animal life.

THE END.

www.ingramcontent.com/pod-product-compliance
Lightning Source LLC
Chambersburg PA
CBHW021946220326
41599CB00012BA/1204